FM 3-06 (FM 90-10)

URBAN OPERATIONS

JUNE 2003

HEADQUARTERS, DEPARTMENT OF THE ARMY

Field Manual
No. 3-06

HEADQUARTERS
DEPARTMENT OF THE ARMY
Washington, DC, 1 June 2003

Urban Operations

Contents

Distribution Restriction: Approved for public release; distribution is unlimited.

*This publication supersedes FM 90-10, 15 August 1979.

Figures

Historical Vignettes

Preface

Doctrine provides a military organization with a common philosophy, a language, a purpose, and unity of effort. To this end, FM 3-06 discusses major Army operations in an urban environment. This environment, consisting of complex terrain, a concentrated population, and an infrastructure of systems, is an operational environment in which Army forces will operate. In the future, it may be the predominant operational environment. Each urban operation will be distinct from any other—any other urban operation as well as similar types of operations in other environments. Each operation will differ because of the multitude of combinations presented by the threat, the urban area itself, the major operation of which it may be part (or the focus), and the fluidity of societal and geo-political considerations. Therefore, there will always exist an innate tension between Army doctrine, the actual context of the urban operation, and future realities. Commanders are responsible to strike the proper balance between preparing for future challenges and maintaining the capability to respond to current threats.

PURPOSE

This manual provides the analytical tools for evaluating an urban operation to determine if the operation is necessary for overall mission success. It also provides the means to understanding and determining the impacts of the urban environment on military operations and provides information on managing, taking advantage of, and mitigating the effects of those impacts as appropriate. As such, this manual demonstrates how to apply the doctrinal principles in FM 3-0 to this unique environment.

SCOPE

Chapter 1 introduces theoretical and historical perspectives of urban operations that serve as the underlying basis for the rest of the manual. Chapter 2 discusses the characteristics of urban centers and populations as well as their impact on operations. It is unlikely that Army forces will ever operate in a benign urban environment; therefore, Chapter 3 discusses the varied nature of potential urban threats. An understanding of the complexities of the urban environment and the nature of the enemy is essential to sound decisionmaking. Chapters 4 and 5 discuss the potential costs of urban operations as well as the effects on each battlefield operating system that the commander and his staff consider early in their planning. These chapters also outline an urban operational framework and specific urban considerations that create the foundations necessary for successfully applying operational doctrine to an urban environment.

The second half of the manual (Chapters 6 – 9) discusses how urban operations are conducted and resourced. Urban operations include major offensive and defensive operations in urban environments as well as stability operations and support operations ranging from peace operations and combatting terrorism to domestic support operations and foreign humanitarian assistance. For the different types of operations—offense, defense, stability, and support—the purpose, characteristics, organization, and considerations are discussed. However, commanders consider that most urban operations will involve some aspect of all four types of operations (although one may dominate) and plan accordingly.

APPLICABILITY

This manual is intended for commanders and their staffs at the brigade through corps level. It addresses the range of operations (both violent and nonviolent) throughout the spectrum of conflict that Army units will execute in urban settings. However, users should also consult JP 3-06 for specific joint information. Additionally, users should be familiar with FM 3-06.11, TC 90-1, and urban operations chapters, appendices, or sections found in other infantry, armor, combined arms, and proponent field manuals for the tactics, techniques, and procedures (TTP) and appropriate proponent information necessary to conduct tactical urban operations at the brigade level and below.

ADMINISTRATIVE INSTRUCTIONS

Chapter 2 defines "city" according to a the population size. However, in historical vignettes and accounts, the term "city" is applied in its common usage without specific regard to size to maintain conformity with most other historical reports.

In this manual, the term "threat" is applied broadly to include an enemy force (conventional or unconventional), an armed belligerent in a peace operation, antagonistic or unfriendly elements of the civilian population, or some other hazardous condition in the urban environment that negatively influences mission accomplishment. The term "hostile" is used as a subset of the threat and denotes a particular element of the urban population (individual, group, or organization) or one or more opposing armed factions in a peacekeeping operation. Both an enemy and a hostile have the *intent* to exploit Army vulnerabilities and negatively affect the urban operation. A hostile, however, is not engaging Army forces in protracted combat operations.

The term military operations on urban terrain (MOUT) is replaced by urban operations (UO). MOUT is an acronym from FM 90-10 *Military Operations on Urban Terrain* that is superseded by this manual.

Otherwise, the glossary lists most terms used in FM 3-06 that have joint or Army definitions. Where Army and joint definitions are different, *(Army)* follows the term. Definitions for which FM 3-06 is the proponent manual (the authority) are marked with an asterisk (*). The proponent or amplifying manual for other terms is listed in parentheses after the definition.

The manual attempts to incorporate historical vignettes into each chapter where the account supports the doctrinal line of reasoning. Two historical vignettes, however, were included as appendices (A and C) because of their longer lengths.

Unless this publication states otherwise, masculine nouns or pronouns do not refer exclusively to men.

The proponent for this publication is HQ TRADOC. Send comments and recommended changes directly to Commander, US Army Combined Arms Center and Fort Leavenworth, Combined Arms Doctrine Directorate, ATTN: ATZL-FD-CD, Futures Development and Integration Center, 1 Reynolds Avenue, Fort Leavenworth, KS 66027-1352.

Chapter 1

Urban Outlook

The ambiguous nature of the operational environment requires Army leaders who are self-aware and adaptive. Self-aware leaders understand their operational environment, can assess their own capabilities, determine their own strengths and weaknesses, and actively learn to overcome their weaknesses. Adaptive leaders must first be self-aware—then have the additional ability to recognize change in their operating environment, identify those changes, and learn how to adapt to succeed in their new environment.

FM 1

Given the prevalence of large cities throughout the world, Army forces, division size and larger, will likely be required to conduct operations in and around large urban areas. These operations will be in support of a joint force commander (JFC) conducting military operations pursuant to United States (US) national security policy. This manual is designed to facilitate the planning and conduct of the full range and spectrum of land operations in a complex urban environment. Each urban environment and urban operation is unique; prescribing specific doctrinal "solutions" for situations is impossible. Instead, this manual provides a framework to commanders and their staffs for understanding the urban environment, for analyzing and deciding whether urban operations (UO) are necessary or feasible, and for applying operational doctrine to this complex environment. It also provides historical vignettes to help develop a refined analytical perspective and some planning points and tactics and techniques to assist in preparing for and conducting UO. Together, this information provides a foundation for approaching major UO, which, combined with other joint and Army doctrine, will help commanders and their staffs learn to adapt and succeed in this challenging environment.

CONTENTS

THE PROSPECT OF URBAN OPERATIONS

1-1. The world is in a period of massive urbanization. A trend of migration from rural to urban areas is occurring throughout the globe. This trend is especially evident in developing nations. Combined with the exponential growth of the global population in the last quarter century, this migration has created massive urban areas that hold the centers of population, government, and economics in their respective regions. In Western Europe, for example, over 50 percent of the land area is urbanized. Just over 30 years ago, only three urban areas in Asia contained at least eight million people. By 2015, estimates show that Asia will have 17 urban areas over ten million, and three of those will top 20 million residents. Almost half of today's population resides in urban areas. Trends also indicate that less developed nations have more centralized societies in a few urban areas. Developed nations spread their centralized societies in several urban areas. In many cases, rapid urbanization has overburdened already weak infrastructures, scarce resources, and a fragile economic base. Given the global population, Army forces will likely conduct operations in and around urban areas—*not as a matter of fate but as a deliberate choice linked to national objectives and strategy and at a time, place, and method of the commander's choosing.*

Army Urban Operations

Army forces conduct UO either as one component of a larger operation or as a single operation focused totally on a specific urban environment. Major Army UO are often part of a joint and multinational effort requiring interagency and civil-military coordination that may include the full spectrum of Army operations. Commanders of Army major operations must determine if UO are essential to mission accomplishment. If so, commanders must carefully integrate the operations into campaign planning to support the operational objectives of the JFC.

Army leaders conducting UO must—

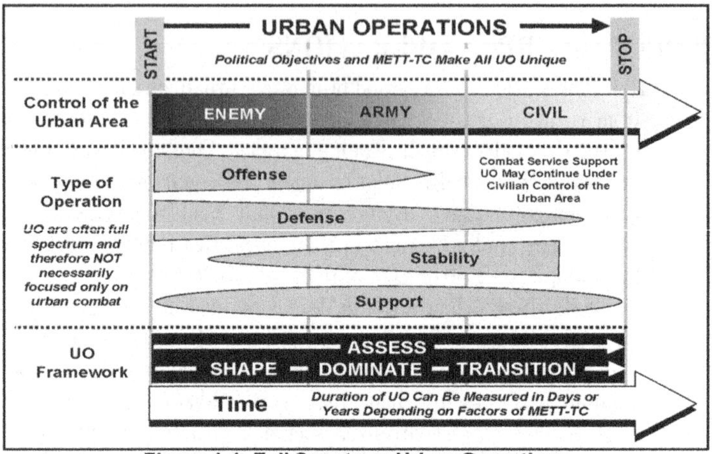

Figure 1-1. Full Spectrum Urban Operations

1-3. When conceptualizing urban operations, commanders understand two important terms: urban area and urban environment. The first is a subset of the second. An ***urban area* is a topographical complex where man-made construction or high population density is the dominant feature.** Focusing on urban areas means concentrating on the physical aspects of the area and their effects on tactics, techniques, and procedures. The ***urban environment* includes the physical aspects of the urban area as well as the complex and dynamic interaction and relationships between its key components—the terrain (natural and man-made), the population, and the supporting infrastructure—as an overlapping and interdependent system of systems.** Critical elements of the infrastructure may lie far beyond the area's physical confines. For example, the generating source providing power to the urban energy system is part of that system but may be located well outside of the urban area. Similarly, effects of the interaction between components of the infrastructure, located both inside and outside the urban area, extend well into smaller, neighboring urban areas and surrounding rural areas and often form their political, economic, and cultural focus. Understanding the total urban environment is essential to planning and conducting the full range of Army urban operations across the spectrum of conflict.

HISTORICAL SIGNIFICANCE OF URBAN AREAS IN WARFARE

1-4. Urban areas always have been central to, or have significantly influenced, military operations. One of the first urban-centered battles was the siege of Troy at the beginning of Greek history. Moreover, much of the history of early Greece revolved around wars between its city-states or with Persia and centered on the conquest, siege, or blockade of cities. Five hundred years later, the Roman Empire replaced Greece as the dominant world power although urban areas remained central to Roman warfare. Even Rome's history can be viewed as a microcosm of urban warfare over the past two thousand years. Though military operations within the physical confines of many of these historic urban areas were not the norm, *the focus* of these operations was their conquest or control.

Rome
A Microcosm of Urban Warfare

During two millennia, Rome has been the center of at least 12 battles. The Gauls lay siege to Rome first in 387 BC. That first siege lasted six months and ended after the barbarians burnt much of the city. The surviving patrician families paid a ransom for the withdrawal of Brennus' army. From 408 to 410 AD, the Goth leader, Alaric, successfully besieged Rome no less than three times. The Byzantine General Belisarius captured Rome twice from the Goths and withstood siege inside the city once between 536 and 549. Five hundred years later in 1084, Norman adventurer Robert Guiscard captured medieval Rome and sacked the city during a dispute between the Pope and the Holy Roman Empire. Forces of the Holy Roman Empire again stormed and captured the city to punish the Pope in 1527. During the Italian Revolution in 1849, a French army supporting the Pope captured the city from the Italian revolutionary army under Garibaldi. In 1944, the last military action took place in and around Rome when the US Fifth Army captured the city from the retreating German army. Rome's turbulent history—fought over ethnic and religious differences, prestige, and military necessity—demonstrates the importance of urban areas in warfare and the various causes and combatants within this complex environment.

1-5. Although Rome last saw combat in 1944, urban areas have been no less prominent in warfare since that time. Beirut in Lebanon, Grozny in Chechnya, and Sarajevo in Bosnia-Herzegovina have been centers of conflict in the last 50 years. Urban areas, now more pervasive than ever before, will continue to be essential to successful operational and strategic warfighting. Today, armies cannot execute major military operations without the influence of surrounding urban environments (with the possible exception of the open desert).

STRATEGIC IMPORTANCE OF URBAN AREAS

1-6. Several reasons have attracted (and continue to attract) armies to combat in urban areas:

1-11. Another reason for engaging in urban operations is the symbolic—historical, cultural, political, and even economic—importance of many urban areas. Often, capital cities—such as Rome, Paris, Seoul, and Berlin—are identified as the strategic centers of gravity of their respective nations. Possessing or threatening these urban areas may impact directly on the outcome of a conflict. The objective of Germany's wars with France in 1870 and 1914 was ultimately Paris. Napoleon's 1812 campaign had as its objective Moscow, as did Hitler's 1941 offensive into Russia. The objective of the Soviet 1945 offensive was Berlin, and the North Vietnamese 1975 offensive had as its objective the South's capital of Saigon. Still, history also reminds us that commanders assess the sustainability and decisiveness of operations directed toward these "prestige" objectives. For example, in 1812, Napoleon captured Moscow but had to evacuate it within 30 days. He lacked supplies and shelter, failed to destroy the Russian Army, and failed to defeat the political will of the Czar and the people. Similarly, the North Korean occupation of Seoul during the Korean War was equally indecisive.

US ARMY'S EXPERIENCE IN URBAN OPERATIONS

1-12. The US Army has a varied history of conducting operations to attack or defend larger urban areas. The American Revolution saw the Army conduct several urban operations. These operations included the unsuccessful defense of New York, the successful attack on Trenton, and the decisive siege and attack on British forces at Yorktown. The Mexican War also had a successful assault on the fortified city of Monterey and the decisive siege of Mexico City. During the American Civil War, the armies, in the tradition of Napoleonic maneuver warfare, avoided urban areas and fought in the open. However, the opposing armies frequently made urban areas their objective because of their importance as railheads. Success in the siege of several key urban areas—Vicksburg, Atlanta, and Petersburg—contributed to the Northern victory.

1-13. Following the Civil War, the US Army faced no large-scale urban combat for several generations. The Indian Wars, the Spanish-American War, the Philippine Insurrection, and even World War I did not require the Army to fight in large urban areas. Between the Civil War and World War II, the US Army fought in several urban areas worldwide supporting US commitments. These limited urban combat operations were small but essential parts of what were urban stability operations. From 1900 to 1901, the Army provided public security for a sector of Peking, China of around 50,000 inhabitants. The Army conducted UO and, in the course of the operation, the 9th US Infantry suffered 20-percent casualties while fighting in Tientsin. Punitive expeditions to places such as Siberia, Cuba, Philippines, Central America, and Mexico put the Army in various urban situations that required using military power, notably, the occupation and security of Vera Cruz, Mexico in 1914. In the context of these smaller-scale contingencies (SSCs), UO became a staple of US Army employment.

1-14. World War II forced the Army to grapple with the issues of large-scale urban combat almost immediately. In his 1941 defense of the Philippines, General MacArthur examined how to defend Manila. Manila represented a large, modern, friendly urban area, which was the capital city of a close US ally. Defending the urban area posed numerous challenges. Ultimately General MacArthur determined that he could best conduct its defense outside the city by defeating the enemy forces in combat on the invasion beaches or shortly after they landed. When Japanese forces defeated MacArthur's Philippine Army in a series of engagements, MacArthur had to decide how best to protect the friendly populace of Manila. He had two choices: abandoning the city or waging a costly defense that would likely result in the city's destruction, thousands of noncombatant casualties, and no operational advantage. He had little choice but to declare Manila an open city and move his forces to Bataan to wage an operational defense in the vain hope that a counteroffensive could relieve his isolated force. On 2 January 1942, Japanese forces entered Manila unopposed.

1-15. Had General MacArthur decided to defend Manila, his forces would have found scant doctrine in the Army regarding how to fight in an urban area. Doctrine for urban operations did not appear until early 1944, when faced with the possibility of fighting through the larger urban areas of Western Europe. At his time the US Army published FM 31-50, *Attack on a Fortified Position and Combat in Towns*. This manual had the first formal discussion of how the Army viewed urban combat. It was based on the Army's limited experiences in the Mediterranean theater and the study of German and Soviet experiences on the Eastern front.

1-16. FM 31-50 emphasized a deliberate pace, individual and small unit initiative, the liberal use of direct and indirect firepower, and decentralized command and execution. It focused on the urban area (as opposed to the environment); however, it did include policies towards the noncombatants. The manual was also focused at the regimental combat team level. Complementing the doctrine of FM 31-50 was the 1944 operations manual, FM 100-5. This latter manual emphasized the importance of combined arms actions and the need for extensive reconnaissance of prepared and defended cities. The Army successfully implemented this doctrine in several major instances of urban combat, most notably the capture of the first German city, Aachen, and hundreds of small-scale urban assaults on cities, towns, and villages across France, the Benelux, and Germany. Army forces also successfully employed this urban combat doctrine during the liberation of Manila in 1945.

1-17. The legacy of this era of Army operations was an effective tactical solution to urban offensive combat: isolate the urban area, seize a foothold, and expand the foothold block by block until occupying the entire urban area and destroying the enemy. The doctrine's emphasis on firepower kept friendly casualties to a minimum. Unfortunately, when enemy forces stoutly defended the urban area, the emphasis on firepower resulted in its virtual destruction and high casualties among noncombatants.

1-18. The doctrinal approach honed in World War II remained the accepted Army approach to urban combat to the century's end. The last successful implementation occurred when liberating Seoul during the Korean War. The Vietnam conflict did not offer the Army opportunities or the requirement to practice urban combat or test and refine doctrine on a large scale. The largest urban battle, Hue, was a chaotic tactical battle that validated most of the historical lessons of urban combat without generating any new doctrinal insights for large-scale urban warfare.

1-19. From the mid-1950s through the 1990s, the Army conducted UO in the United States in support of civil authorities during civil unrest and anti-Vietnam protests. Some operations involved numerous active and reserve component forces engaged in restoring public order. The Detroit riots of 1967 and the Los Angeles riots of 1992 required the commitments of active and National Guard units. In 1968, the Army deployed over 35,000 troops to Washington D.C., Chicago, and Baltimore following the death of Dr. Martin Luther King, Jr.

1-20. In the 1970s and 1980s, Army doctrine predominantly focused on urban areas and successfully fighting a conventional ground war against Soviet and Warsaw Pact forces in Central Europe. The 1979 FM 90-10, *Military Operations on Urbanized Terrain (MOUT),* described how to conduct urban operations against Soviet forces in Germany. Its concepts were never tested other than in simulation, and its approach to urban combat was not substantially different from that practiced by the Army since World War II. Despite previous doctrine's admonition to avoid cities, the Army has had to fight in them in diverse circumstances.

MODERN ARMY URBAN OPERATIONS

1-21. Modern urban operations span the full range of possible applications of military power. At the high end of the spectrum of conflict is major theater war (MTW) dominated by offensive and defensive operations that, when undertaken, will commonly include urban operations. At the lowest level are a multitude of urban peacetime military engagement (PME) activities. These activities foster and strengthen alliances and coalitions as well as deter aggression on the part of potential threats. At mid-level between MTW and PME are SSC urban operations. As a result of being mid-range, any type of operation may potentially dominate an SSC; however, the various urban stability operations form the majority. At higher echelons, these separations are often viewed as levels of intensity. For the tactical units conducting urban operations, these divisions appear indistinct, as the intensity is often high despite where the operation falls within the level of conflict.

MAJOR THEATER WAR

1-22. While UO in a MTW can encompass the full range of Army operations, the offense and defense will be central and decisive to success. Although mindful of collateral damage and noncombatants, urban operations in a MTW (compared to urban operations in SSCs or as part of PME activities) will be the least constrained because vital national interests will be at stake. UO in a MTW, therefore, will require a significant investment of resources of all types. Specialized units such as psychological operations, civil affairs, and other special operations forces (SOF) will likely be in high demand. UO in a MTW will require an abundance of infantry and may require significant casualty replacements and medical support. Logistics to support the distinctive urban environment includes large amounts of lethal and nonlethal specialty munitions, such as smoke, precision field artillery rounds, demolitions, and hand grenades.

1-23. Of potential urban scenarios confronting the future Army, urban offensive and defensive operations in an MTW are the most dangerous and challenging. They will take one of two principal forms: fluid or siege. In a fluid urban combat operation, both sides may contend for position and advantage in the urban battlespace. The attacker will seek to quickly seize decisive points before the enemy is able to establish a cohesive defense. This will likely require the attacker to bypass enemy defensive positions whose occupation or reduction are not critical to mission success. Conversely, the defender may use interior lines to shift forces in a fluid defense. In a siege, one side clearly has the initiative as the attacker, and the other side has the advantages of the defense. A siege situation can develop as a result of an initial fluid urban battle, or it may be a function of previous military operations that occurred outside the urban area. The Army doctrine's emphasis on initiative, agility, depth, synchronization, and versatility generally supports the fluid form of urban combat; however, commanders also understand that the factors of METT-TC may support a longer-term, siege approach.

SMALLER-SCALE CONTINGENCIES

1-24. SSCs encompass a wide range of military operations that fall between MTW and PME and frequently involve urban operations. SSCs are conducted to facilitate diplomacy and support political initiatives, protect American lives and interests, and disrupt illegal activities. Joint task forces (JTFs) typically conduct SSCs although one service may provide the bulk of the force. During these urban contingencies, resources are often more limited and the restraints on applying combat power are greater as the need to maintain legitimacy will grow in importance. Typically, Army forces will need the assistance of multinational partners, other agencies, local noncombatants, and nongovernmental organizations (NGOs) to successfully complete the mission.

PEACETIME MILITARY ENGAGEMENTS

1-25. UO, at the lowest level of conflict, may also take many forms. They serve to strengthen alliances and coalitions, discourage arms races, combat terrorism, and generally reduce the potential for instability and conflict. Combat in PME activities is not the norm. They are least likely to involve the use of force (when necessary, nonlethal is preferred). The presence of Army forces performing PME activities in foreign urban areas provides a visible sign of US commitment to peace and stability in that region. In many of these lower-intensity UO, Army forces often support other agencies. These other agencies actually plan and lead the operation. Army forces provide military capabilities (to include organization and leadership), manpower, equipment, and other resources not readily available. As with UO in SSCs, proactive and aggressive interaction and coordination with multinational partners, governmental and nongovernmental organizations, and the urban populace will be vital to success.

PREPARING FOR FUTURE URBAN OPERATIONS

1-26. To operate successfully in a complex urban environment requires rigorous, realistic UO training. Training is conducted by the complete combined arms team and covers the full range of Army operations. It also replicates—

Chapter 2

Urban Environment

From a planning perspective, commanders view cities not just as a topographic feature but as dynamic entities that include hostile forces, local population, and infrastructure. Planning for urban operations requires careful IPB, with particular emphasis on the three-dimensional nature of the topography and the intricate social structure of the population.

FM 3-0

Of all the environments in which to conduct operations, the urban environment confronts Army commanders with a combination of difficulties rarely found elsewhere. Its distinct characteristics result from an intricate topography and high population density. The topography's complexity stems from the man-made features and supporting infrastructure superimposed on the natural terrain. Hundreds, thousands, or millions of civilians may be near or intermingled with soldiers—friendly and enemy. This second factor, and the human dimension it represents, is potentially the most important and perplexing for commanders and their staffs to understand and evaluate. The intelligence preparation of the battlefield (IPB) *process* remains unaffected by urban areas (see FM 34-130 and Appendix B); this chapter provides information essential to the conduct of the IPB for an urban environment.

Although urban areas possess general similarities, each environment is distinct and will react to and affect the presence and operations of Army forces differently. A tactical technique effective in one area may not be effective in another area due to physical differences, such as street

CONTENTS

patterns or the type of building construction. An Army policy popular with one urban group may cause resentment and hostility in another due to diverse cultural differences. All difficulties potentially exist, but they increase the complexity for Army forces operating in urban areas. These difficulties range from conventional military forces to disease and starvation (see Chapter 3) to a pervasive media—often acutely present in intricate combinations. Thus, commanders at all levels make extraordinary efforts to assess and understand their particular urban environment to plan, prepare for, and execute effective urban operations (UO).

A COMPLEX ENVIRONMENT

2-1. Urban areas vary depending on their history, the cultures of their inhabitants, their economic development, the local climate, available building materials, and many other factors. This variety exists not only among urban areas but also within any particular area. The ever-changing mix of natural and man-made features in urban areas present commanders with some of the most difficult terrain in which to conduct military operations.

2-2. Although urban areas possess similar characteristics, no two are identical. The sprawl of Los Angeles, for example, bears little physical resemblance to New Delhi. Societal characteristics most significantly affect each area's uniqueness and complexity. While complex, information about the terrain, its potential effects on operations, and how it changes over time may be determined with some degree of certainty. However, the human dimension is much more difficult to understand and assess, particularly its effects on military operations. Like any environment, *the side that can best understand and exploit the effects of the urban environment has the best chance of success.*

2-3. Whether a large metropolis or a small village, each urban environment has an identifiable system of components that constantly change and interact. This "system of systems" consists of the *terrain*, the *society*, and the *infrastructure* that links the two (see Figure 2-1). (These categories highlight the key aspects to understanding the urban environment and will be used throughout the manual; however, the civil-military operations (CMO) discussion in Chapter 9 provides an alternate method for categorizing and assessing the effects of civil considerations in *any* operational environment.)

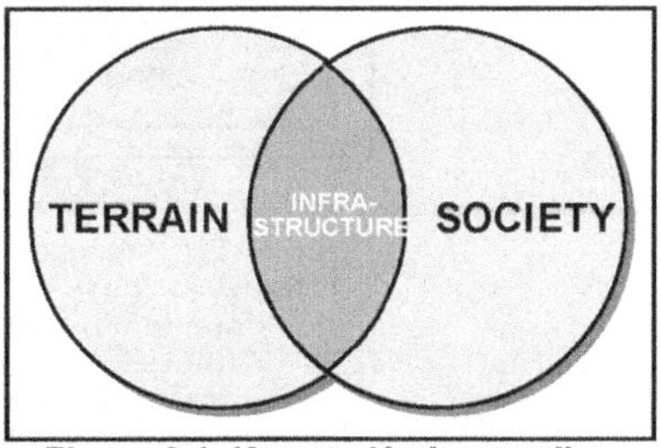

Figure 2-1. Keys to Understanding the Urban Environment

2-4. These systems are not separate and distinct categories but rather overlapping and interdependent. Thoroughly analyzing these elements, along with the other factors of mission, enemy, weather, troops and support available, time, and civil considerations—

cause (either intentionally or not) uncontrollable fires or the loss of electricity. A power outage can cause flooding (especially in subsurface areas) by shutting down pumping stations. Entire buildings may be destroyed, eliminating reference points and leaving large piles of rubble. Additionally, buildings and other urban structures, damaged but not destroyed, can still be effective obstacles and possible booby traps. Their weakened construction and unstable structure increase the risk of injury to soldiers and civilians moving within them. (Engineers often determine whether the buildings can support occupation by Army forces or civilians.) The likely presence of toxic industrial materials (TIM) can create additional obstacles.

2-10. Commanders in other environments normally address the depth, breadth, and height of their AO in terms of two areas: *airspace* and *surface*. In an urban environment, they broaden their scope to include *supersurface* and *subsurface* areas (see Figure 2-2). Although spatially separated, each area may be used as an avenue of approach or mobility corridor, line of communications (LOC), and engagement area.

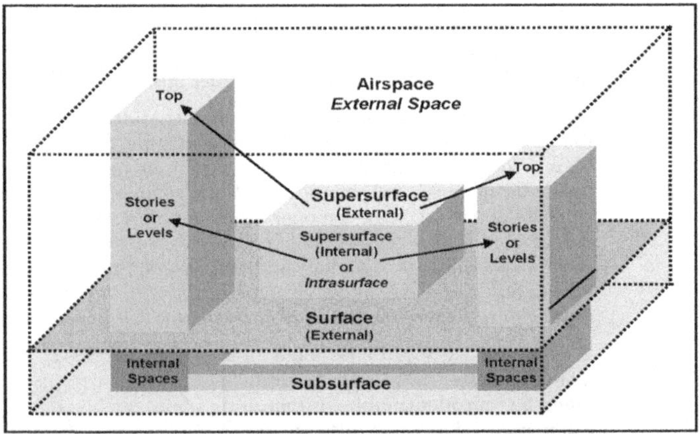

Figure 2-2. The Multidimensional Urban Battlefield

2-11. Supersurface and subsurface areas magnify the complexity of the urban physical environment. Commanders consider activities that occur outside buildings and subterranean areas (the *external* space) as well as the activities that occur unseen in buildings and subterranean systems (the *internal* space). The internal space further challenges command, control, and intelligence collection activities and increases the combat power required to conduct UO. Commanders develop methods to help themselves, their staffs, and their subordinate commanders and staffs to represent and visualize the multiple dimensions. Such dimensions can change rapidly simply due to continued urban growth or, as described earlier, the effects of nature and UO themselves.

Airspace

2-12. Aircraft and aerial munitions use the airspace as rapid avenues of approach in urbanized areas. Forces can use aviation assets for observation and reconnaissance, aerial attack, or high-speed insertion and extraction of soldiers, supplies, and equipment. Some surface obstacles, such as rubble, do not affect aviation assets. However, buildings of varying height and the increased density of towers, signs, power lines, and other urban constructions create obstacles to flight and the trajectory of many munitions (masking). These obstacles can limit low-altitude maneuverability in the urban airspace. Excellent cover and concealment afforded enemy gunners in an urban area increases aviation vulnerability to small arms and man-portable air defense systems (MANPADS), particularly when supporting ground forces.

Surface

2-13. Surface areas apply to exterior ground level areas, such as parking lots, airfields, highways, streets, sidewalks, fields, and parks. They often provide primary avenues of approach and the means for rapid advance. However, buildings and other structures often canalize forces moving along them. As such, obstacles on urban surface areas usually have more effect than those in open terrain since bypass often requires entering and transiting buildings or radical changes to selected routes. Where urban areas abut the ocean or sea, large lakes, and major rivers, the surface of these bodies of water may provide key friendly and threat avenues of approach or essential LOCs and, therefore, may be a significant consideration for Army commanders. As such, amphibious and river-crossing operations may be an integral part of the overall urban operation.

2-14. Larger open areas—such as stadiums, sports fields, school playgrounds, and parking lots—are often critical areas during urban operations. They can provide locations for displaced civilians, interrogation centers, and prisoner of war holding facilities. These areas also can afford suitable aircraft landing and pickup zones and artillery firing locations. They can provide logistic support areas and aerial resupply possibilities because they are often centrally located.

Supersurface

2-15. These areas include the internal floors or levels (*intrasurface* areas) and external roofs or tops of buildings, stadiums, towers, or other vertical structures. They can provide cover and concealment; limit or enhance observation and fields of fire; and restrict, canalize, or block movement. However, forces can move within and between intrasurface areas creating additional, though normally secondary, avenues of approach. Rooftops may offer ideal locations for landing helicopters for small-scale air assaults and aerial resupply. First, engineers analyze buildings for their structural integrity and obstacles. Such obstacles include electrical wires, antennas, and enemy-emplaced mines (although personnel may be inserted by jumping, rappelling, or fast roping from a hovering helicopter and extracted by hoist mechanisms). Some rooftops are designed as helipads. Roofs and intrasurface areas may also provide excellent locations for snipers; lightweight, handheld antitank weapons; and MANPADS. They enable top-down attacks against the weakest points of armored vehicles and unsuspecting aircraft. Overall, elevated firing positions reduce the value of any cover in surrounding open areas and permit engagement at close range without risk of immediate close assault. This area (and the subsurface area) requires commanders to think, plan, and execute ground operations vertically as well as horizontally. In this latter regard, UO share strong similarities with mountain operations (see FM 3-97.6).

Subsurface

2-16. These areas are ***subterranean*** or below surface level. They may serve as secondary and, in fewer instances, primary avenues of approach at lower tactical levels. When thoroughly reconnoitered and controlled, they offer excellent covered and concealed LOCs for moving supplies and evacuating casualties. They may also provide sites for caching and stockpiling supplies. Subsurface areas include the subways, tunnels, sewers, drainage systems, cellars, civil defense shelters, and other various underground utility systems. In older cities, they may include ancient hand-dug tunnels and catacombs. Both attacker and defender can use subsurface areas to gain surprise and maneuver against the rear and flanks of a threat and to conduct ambushes. However, these areas are often the most restrictive and easiest to defend or block. Their effectiveness depends on superior knowledge of their existence and overall design. Army commanders may need to consider potential avenues of approach afforded by the subsurface areas of rivers and major bodies of water that border urban areas. This particularly applies when operating as part of a joint task force (JTF) task organized with SOF or when opposing a threat with similar capabilities.

BROAD URBAN PATTERNS

2-17. Four major urban patterns can influence UO (see Figure 2-3). Central to two of the patterns (satellite and network) is the ***hub*** or dominant urban area or pattern around which outlying urban areas or patterns radiate. (A segmented urban area, because it tends to be a larger urban area, can often be a hub.) In offensive and defensive operations, the hub serves as a pivot or strong point; as such, it often becomes a major obstacle to an attacker. If the attacker chooses to bypass the urban area (hub) located along his axis of advance without first isolating the area, he may expose his flank to attack from the hub as well as dependent urban areas or subordinate satellite patterns. Because the focus of stability operations and support operations is on people, commanders understand the value and influence of the hub to the economic, political, or cultural well being of the surrounding area. Whether or not a hub, commanders must remember that urban areas are not islands; all are connected to the surrounding rural (and other urban) areas through fluid and permeable boundaries and LOCs.

Satellite Pattern

2-18. This common pattern consists of a central hub surrounded by smaller, dependent urban areas. LOCs tend to converge on the hub. The natural terrain throughout this pattern is relatively homogenous. Outlying areas often support the principal urban area at the hub with means of reinforcement, resupply, and evacuation. In some instances, they may serve as mutually supporting battle positions. Commanders should consider the effects of the outlying urban areas on operations within the hub, and, conversely, the effects of operations within the hub on outlying urban areas. Information operations (IO), for example, targeted primarily at the hub of a satellite pattern may subsequently influence outlying urban areas and achieve necessary effects without having to commit specific resources to these areas.

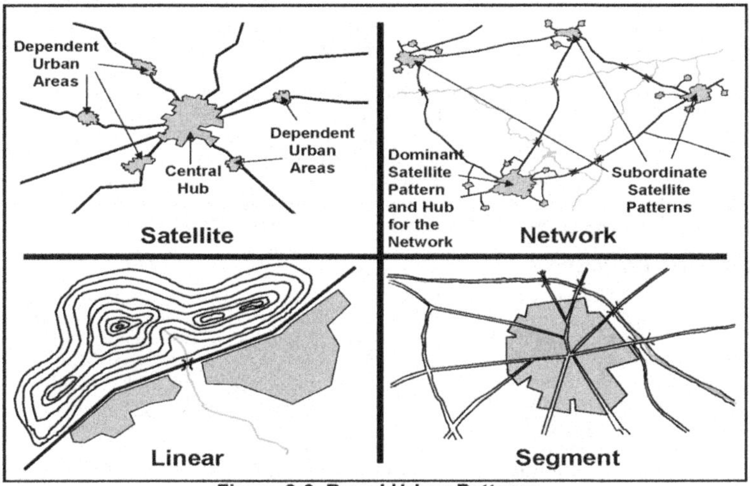

Figure 2-3. Broad Urban Patterns

Network Pattern

2-19. The network pattern represents the interlocking of the primary hubs of subordinate satellite patterns. Its elements are more self-sufficient and less supportive of each other, although a dominant hub may exist. Major LOCs in a network extend more than in a satellite pattern and take more of a rectangular rather than a convergent form. Its natural terrain may vary more than in a single satellite array. Operations in one area may or may not easily influence, or be influenced by, other urban areas in the pattern.

Linear Pattern

2-20. Potentially a subelement of the previous two patterns, the linear pattern may form one ray of the satellite pattern or be found along connecting links between the hubs of a network. Most frequently, this pattern results from the stringing of minor urban areas along a confined natural terrain corridor, such as an elongated valley, a body of water, or a man-made communications route. In offensive and defensive operations, this latter form of the linear pattern facilitates developing a series of strong defensive positions in depth, effectively blocking or delaying an attacking force moving along the canalized terrain.

Segment Pattern

2-21. When dominant natural terrain, such as a river or man-made features (canals, major highways, or railways), divides an urban area, it creates a segmented pattern. This pattern often makes it easier for commanders to assign areas of operations to subordinate commanders. However, this pattern may fragment operations and increase risk to an operation requiring mutual support between subordinate units. Still, the segmented urban areas may allow commanders to isolate threats more easily in these areas and focus operations within segments that contain their decisive points. Although an integral part of the whole (the urban area), each segment may develop distinct social, economic, cultural, and political characteristics. This *social segmenting* may benefit commanders faced with limited assets to influence or control the urban populace. After thoroughly analyzing the society, they may be able to focus IO and populace and resources control measures against only specific segments that affect decisive operations. Commanders may need only to isolate other segments or may need to just monitor for any significant changes in the attitudes, beliefs, or actions of the civilians located there.

LESSER STREET PATTERNS

2-22. Lesser patterns in the urban area result from the layout of the streets, roads, highways, and other thoroughfares. They evolve from influences of natural terrain, the original designer's personal prejudices, and the changing needs of the inhabitants. Street patterns (and widths) influence all BOS; however, they greatly affect maneuver, command and control, and combat service support. (In some portions of older Middle Eastern urban areas, the labyrinths of streets were designed only to allow two loaded donkeys to pass each other; tanks are too wide.) Urban areas can display any of three basic patterns and their combinations: *radial*, *grid*, and *irregular* (see Figure 2-4).

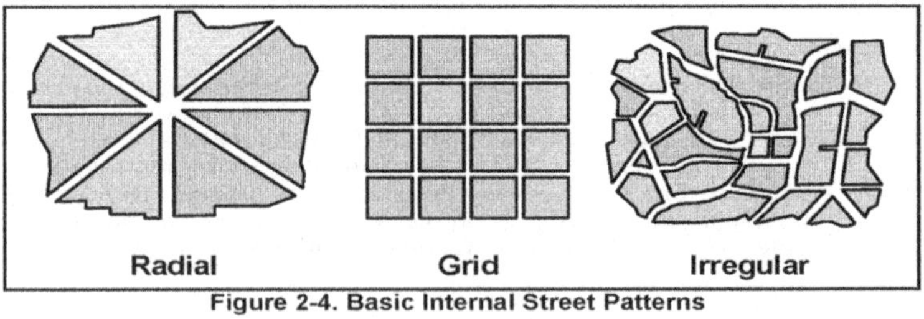

Figure 2-4. Basic Internal Street Patterns

Radial

2-23. Societies of highly concentrated religious or secular power often construct urban areas with a radial design: all primary thoroughfares radiating out from the center of power. Cities with this design may signal an important historical aspect in the overall analysis of the urban society. Terrain permitting, these streets may extend outward in a complete circle or may form a semicircle or arc when a focal point abuts a natural barrier, such as a coastline or mountain. To increase mobility and traffic flow, societies often add concentric loops or rings to larger radial patterns. Unless commanders carefully plan boundaries, routes, and axes of advance, their subordinate units' movement or maneuver may be inadvertently funneled toward the center of urban areas with this pattern resulting in congestion, loss of momentum, and an increased potential for ambush or fratricide.

Grid

2-24. The most adaptable and universal form for urban areas is the grid pattern: lines of streets at right angles to one another forming blocks similar to the pattern of a chessboard. A grid pattern can fill in and eventually take over an original radial pattern. Grid patterns often appear to ease the assignment of boundaries for subordinates units. However, commanders also consider how the natural terrain influences operations and the establishment of control measures. They also consider the influence of the buildings and other structures lining these streets, such as their height and construction, before assigning boundaries and developing other control measures.

2-25. Describing boundaries and phase lines by easily recognizable features is as important in urban areas as elsewhere. If available, natural features are a better descriptor than man-made features that may be altered or unrecognizable. When Army forces work closely with local law enforcement agencies, commanders may not need to assess the effect of street patterns on the assignment of boundaries. Instead, commanders may assign boundaries overlaid on existing administrative boundaries used by local law enforcement agencies to increase interoperability and aid in unity of effort.

Irregular

2-26. In most urban areas, regardless of the original intent, plan, or vision, existing street patterns emerge from successive plans overlaid one on another. Some are well planned to fit with previous plans while others a haphazard response to explosive urban growth. The result may mix patterns. Urban engineers and planners may specifically design irregular patterns for aesthetic reasons (as in many suburban housing developments) or to conform to marked terrain relief. Irregular street patterns may alert commanders and analysts that the underlying natural terrain may exert greater influence over operations than in other portions of the urban area. Finally, irregular street patterns make the movement and maneuver of forces less predictable.

AN URBAN MODEL

2-27. Throughout the world, urban areas have similar form and function. In form, urban areas contain like characteristics, readily divisible into distinct sections or areas. Functionally, they tend to be the centers of population, finance, politics, transportation, industry, and culture. While urban areas may be modeled by several different means, Figure 2-5 on page 2-10 illustrates the general forms and internal functions. Some forms and functions may overlap. For example, high-rise buildings are located in core areas as well as in outlying areas and may be used for residential purposes. With the rapid urbanization associated with developing nations, the areas displayed in this urban model often manifest themselves less clearly there than in developed nations.

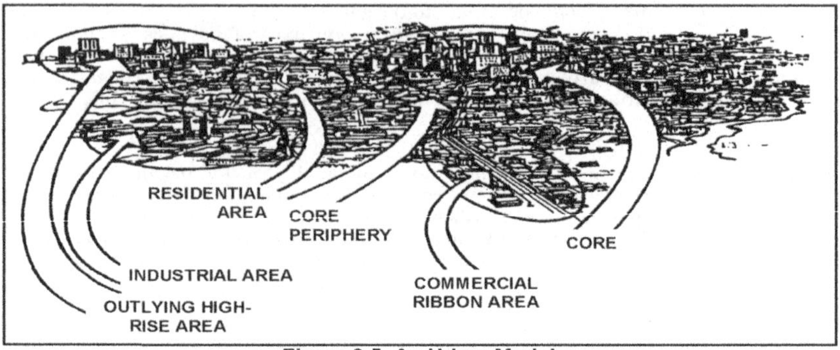

Figure 2-5. An Urban Model

2-28. This analysis helps to determine, in general terms, potential advantages and disadvantages each portion of the urban area may have toward accomplishing the urban operation. However, construction materials and methods can vary drastically. Commanders identify specific building types and construction and understand weapons effects on them. If a commander desires precise effects, the chosen munitions or weapons system must be sufficiently accurate, capable of penetrating the target structure (without exiting the other side), and achieve effects within. Often noncombatants, critical infrastructure, or protected targets are in the vicinity. Commanders may need to determine if the surrounding walls or structures will sufficiently absorb or negate the blast or thermal effects of the weapon. Regardless, understanding the structure of buildings in the urban AO allows commanders to determine the best means to accomplish the mission.

Core

2-29. The core is the heart of the urban area, the downtown or central business district. Relatively small and compact, it contains a large percentage of the urban area's shops, offices, and public institutions. Often, it houses the headquarters for commercial and financial activities and contains important cultural, historical, and governmental buildings. These activities prefer the core because of its accessibility. As the focal point of the transportation network, residents find the core the easiest part of the urban area to reach. It normally has the densest concentration of multistory buildings and subterranean features (underground parking garages, underground shopping centers, and basements).

2-30. High-rise buildings, varying greatly in height (possibly 50 stories above ground and four stories below ground), make up the cores of today's urban areas. Buildings routinely abut one another, with little or no setback from the sidewalks. Building height and density (except in outlying high-rise areas) often decreases from the core to the edge of the residential areas, while the amount of open areas frequently increases. Modern urban planning allows for more open spaces between buildings than found in the cores of older urban areas. Most core areas have undergone constant redevelopment resulting in various types of construction. Commonly, brick buildings abound in the oldest part of the core; framed, heavy-clad structures in the next oldest part; and a concentration of framed, light-clad buildings in the newest part. The outer edge of the core, the core periphery, has ordinarily undergone less change than the core resulting in buildings of uniform height (commonly two to three stories in towns and five to ten stories in larger urban areas).

2-31. Generally, offensive operations focused in core areas (even when effectively isolated) will require greater resources—particularly manpower, time, and information—than in many other parts of the urban area. Mounted maneuver often proves more difficult in core areas because of fewer open areas, buildings closer to the streets, and more civilian vehicles. Rubbled buildings in central core areas (especially high-rise buildings) become greater obstacles to mobility as they can collapse on and easily block thoroughfares. Rubble piles can afford excellent covered and concealed positions for dismounted threat forces. Consequently, commanders use more dismounted forces as part of their combined arms operations. Conversely, the core may be critical to urban defensive operations, particularly older areas of heavier construction that afford greater protection. Despite potential difficulties, the core area may be key to accomplishing many stability or support missions since it houses much of the human activity that occurs in the urban area.

Industrial Area

2-32. Industrial areas often develop on the outskirts of the urban areas where commercial transportation is easiest (along airfields and major sea, river, rail, and highway routes). These areas will likely displace farther from the core and residential areas as urban planners recognize the potential threat of TIM. The dispersed pattern of the buildings provides sufficient space for large cargoes, trucks, and materiel handling equipment. These areas may provide ideal sites for logistic bases and maintenance sites. While older, heavier-clad structures may be found, new construction consists of low, large, flat-roofed factory and warehouse buildings with large parking areas and work yards. These structures generally have steel frame and lightweight exterior walls. Multistory structures usually have reinforced concrete floors and ceilings.

2-33. Toxic industrial chemicals and other TIM may be transported through an urban area (by rail, barge, truck, or pipeline) or found stored throughout. However, larger concentrations will exist in industrial areas, and their presence should concern Army forces operating near them.

2-34. Each year, over 70,000 different chemicals are produced, processed, or consumed globally. An estimated 25,000 commercial facilities around the world produce, process, or store chemicals that have a legitimate industrial use yet are also classified as chemical warfare agents. Many other chemicals (not classified as weapons) may still be sufficiently hazardous to pose a considerable threat to Army forces and civilians in urban areas as choking agents or asphyxiates, flammables or incendiaries, water contaminants, low-grade blister or nerve agents, or debilitating irritants. These chemicals can be released either accidentally or deliberately. On 2 December 1984, nearly 40 tons of methylisocyanate used to produce pesticides leaked from a storage tank at Bhopal, India, killing thousands and injuring hundreds of thousands. Figure 2-6 contains a small sampling of other toxic industrial chemicals along with their industrial or commercial usage that commanders may encounter in an urban area. The most common chemicals that pose a risk to Army forces are highly toxic irritant gases such as ammonia, chlorine, hydrogen chloride, and sulfur dioxide.

Toxic Industrial Chemical	Industrial/Commercial Uses
Ammonia	Commercial Refrigerant, Fertilizer and Food Production, Petroleum, Explosives, Other Chemicals
Arsine	Semiconductor Industry
Boron Trichloride	Organic Catalyst, Soldering Magnesium
Boron Trifluoride	Chemical Catalyst, Aluminum Refining
Carbon Disulfide	Industrial Solvent, Dry-Cleaning, Agriculture, Petroleum, Electroplating
Chlorine	Potable Water, Disinfectants, Metal Treatment, Plastics & Rubber
Diborane	Plastics and Rubber
Ethylene Oxide	Industrial Alcohols, Fumigant, Industrial Sterilant
Fluorine	Uranium Processing, Rocket Fuel
Formaldehyde	Plastics, Fertilizers, Preservative/ Corrosion Inhibitor, Fungicide and Germicide, Pesticide, Pharmaceuticals
Fuming Nitric Acid	Fertilizers, Explosives, Metal Processing, Pesticides, Rocket Fuel
Hydrogen Bromide	Chemical Industry, Pharmaceuticals
Hydrogen Chloride	Fabrics, Semiconductors
Hydrogen Cyanide	Pesticides, Other Chemicals, Pharmaceuticals, Electroplating
Hydrogen Fluoride	Glass Production, Chemical Catalyst
Hydrogen Sulfide	Metallurgy, Agricultural Disinfectant
Phosgene	Dyes, Pharmaceuticals, Herbicides & Insecticides
Phosphorus Trichloride	Metallurgy, Pesticides and Germicides, Gasoline Additive
Sulfur Dioxide	Paper, Food Processing, Ice Production, Disinfectant, Leather Processing
Sulfuric Acid	Fertilizers, Petroleum, Iron and Steel Production, Battery Electrolyte
Tungsten	Electronics, Other Chemicals

Figure 2-6. Toxic Industrial Chemicals and Their Industrial or Commercial Uses

2-35. Standard chemical defense equipment may not protect against (and chemical detection devices may fail to detect) many toxic industrial chemicals. Therefore, the risk to soldiers operating near the chemicals may increase. Commanders vigilantly identify these potential hazards, carefully consider them as part of their overall vulnerability analysis, factor the analysis into their risk assessment, and execute necessary contamination avoidance measures. Any assessment includes the chance that toxic industrial chemicals may be deliberately released by a threat to gain advantage or accidentally released by friendly actions (see FM 3-21 and FM 3-14).

Outlying High-Rise Area

2-36. High-rise areas consist of multistoried apartments, commercial offices, and businesses separated by large open areas, such as parking lots, parks, and individual one-story buildings. High-rise buildings are framed, light-clad construction with thin walls of brick, lightweight concrete, or glass. The automobile, mass transit systems, and improved road networks encourage these areas to grow and function further from the urban core.

2-37. Similar to the urban core, units given the mission to clear these areas, or even portions therein, will need more resources—most notably personnel and time—to accomplish their mission. Commanders should consider courses of action that isolate these entire areas, multiple sections within these areas, or even individual buildings before assigning tasks. The tasks could rapidly drain a unit's resources or unhinge other portions of the major operation. When defending, commanders who can integrate these areas in the defense will present the attacker with similar resource problems and may be appropriate in a defense to delay. However, defending commanders ensure that the defense is arranged so that this portion cannot be easily isolated and bypassed. Defensive positions in structures may require extensive reinforcement due to light-clad construction.

Residential Area

2-38. Residential areas can be found dispersed throughout the urban area; however, large suburban areas (or sprawl) normally form on the outskirts. Residential areas often consist of row houses or single-family dwellings set in a grid or ringed pattern in a planned development project. Yards, gardens, trees, and fences usually separate the buildings in a residential area. Modern residential construction is often of light-clad, framed wood construction, or brick. The combined population of surrounding suburban areas often far outnumbers that of the urban area proper. Specific suburbs tend toward homogeneity based on ethnicity, religion, economics, or some other social aspect. Commanders locate and analyze these areas to determine their impact on operations—often the most critical importance is the people located there (see the subsequent discussion in this chapter on the urban society).

2-39. In offensive and defensive operations, commanders determine whether operations pose an unacceptable physical risk to civilians. If so, they may have to relocate civilians to a safer area, perhaps another residential area. If not, commanders may implement a "stay-put" policy for that area and attempt to isolate the effects of the operation from them. During support operations, residential locations may be the initial focal point for operations since most of the permanent population is located there.

2-40. This area also contains a relatively recent urban phenomenon known as *shantytowns*. These areas are commonly on unoccupied, low-value land in and around many urban areas in underdeveloped countries. Shantytowns may contain over 50 percent of the total urban population. They usually lack streets and public utilities. The lean-to structures tend to be irregularly laid out, connected by walking paths, and made of any scrap material available: lumber, brick, sheet metal, cardboard, cloth, or vegetation. The random arrangement of structures, the absence of formal street naming and numbering, and often the lack of easily identifiable buildings and terrain create challenges. These challenges include navigating, coordinating, and transmitting accurate information and intelligence. Depending on the operation, the temporary nature of the structures can also mean that mobility can be either more or less restricted than other sections of the urban area. A military force may easily knock down and traverse structures without affecting mobility at all. However, their destruction may cause unacceptable civilian casualties, in which case mobility becomes more restrictive as the narrow paths often do not accommodate vehicular traffic. Similarly, the makeshift materials inhibit weapons effects less than many other parts of the urban area built more solidly. A tank round, for example, may go much farther and injure many more noncombatants than in an area where the primary building material is stone. Regardless, commanders consider the effects of their operations in this area, to include vehicles and weapons, as the weak structures increase the risk of fratricide, civilian casualties, and large, rapidly spreading fires.

Commercial Ribbon Area

2-41. Commercial ribbon areas are rows of stores, shops, and restaurants built along both sides of major streets that run through and between urban areas. These same types of areas often develop along the roads that connect one urban area to another (strip areas). The buildings uniformly stand two to three stories tall (about one story taller than the dwellings on the streets behind them).

URBAN SOCIETY

2-42. Although intricate, understanding the urban terrain is relatively straightforward in comparison to comprehending the multifaceted nature of urban society. UO often require Army forces to operate in close proximity to a high density of civilians. Even evacuated areas can have a stay-behind population in the tens of thousands. This population's presence, attitudes, actions, communications with the media, and needs may affect the conduct of operations. Homogeneity decreases drastically as the size of the urban area increases. Commanders take into account the characteristics of a population whose beliefs and interests vary based on factors. Figure 2-7 lists the factors. Civilian populations continually influence, to varying degrees, operations conducted in an urban area. Thoroughly understanding these societal aspects and avoiding "mirror-imaging"—overlaying one's own values and thought processes on top of the person or group one is trying to assess—will help to accurately anticipate civilian actions and response.

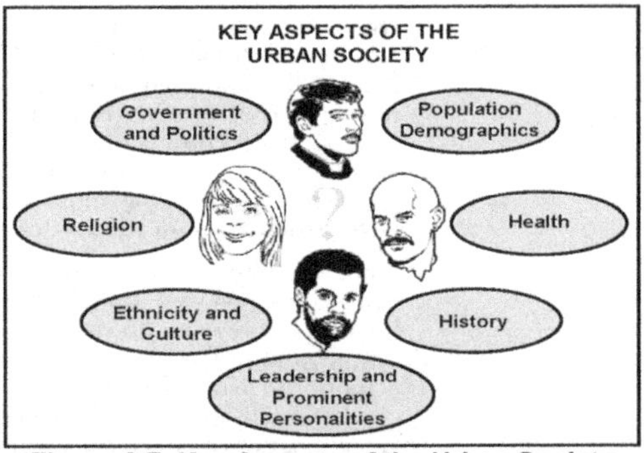

Figure 2-7. Key Aspects of the Urban Society

POTENTIAL CENTER OF GRAVITY

2-43. A COG during an urban operation, particularly in stability operations and support operations, may be the civilian inhabitants themselves—specifically their behavior. However, supportive behavior is generally an advantage in any type of operation. Correspondingly, neutral behavior toward friendly forces is an advantage over hostile behavior. To influence or control their behavior, commanders first understand the society's complex nature and character. Second, they understand and accept that every military action (or inaction) may influence the relationship between the urban population and Army forces, and, by extension, mission success. Lastly, they understand that Army forces may play only a supporting (but essential) role as part of an integrated and synchronized multiagency effort focusing all aspects of national power. With this awareness, commanders can take one or more actions:

divisions do exist (either physical or ideological), commanders can more easily determine those divisions and their fundamental causes with smaller populations.

2-47. As urban areas expand, the urban patterns begin to blur and the social complexity increases. For example, as satellite patterns continue to grow, the LOCs between a central hub and outlying urban areas may develop and begin to assume a linear urban pattern. Simultaneously, a hub and outlying urban areas may continue to expand until they merge into a single, large metropolis. On a larger scale, a network pattern can grow and unite as a single megalopolis. This growth physically unites smaller urban areas but cannot force conformity of needs and beliefs. It also increases the physical and social complexity of an urban area.

GROUP SIZE, LOCATION, AND COMPOSITION

2-48. Understanding how specific elements of the urban society affect operations (and vice versa) normally begins with analyzing their *size, location,* and *composition* (see Figure 2-9). Because commanders must minimize civilian casualties, size and location (without regard to composition) are important initial demographic considerations. After determining the presence and numbers of civilians relative to decisive points, commanders can then decide whether civilian *proximity* and *density* represent a significant risk to their mission—refugees clogging LOCs, for example. If civilians are the primary focus of the operation, as in many stability operations and support operations, this same analysis may help to determine decisive points. In this analysis, commanders consider that urban areas, on many levels, are in constant motion. The densities of circulating people and other traffic often vary according to the time of day, such as rush hours and market times. In planning urban operations, commanders may need to consider the timing or rhythms of population and vehicular movements in the urban area.

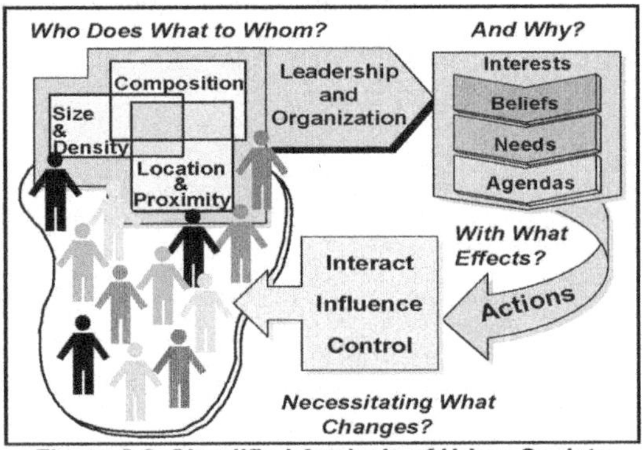

Figure 2-9. Simplified Analysis of Urban Society

2-49. Commanders determine the composition of, or the identifiable groups or organizations within, the civilian urban population. Groups may be categorized by race, religion, national origin, tribe, clan, economic or social class, party affiliation, education level, union memberships, age, gender, occupation, or any other significant social grouping. Physical and ideological overlaps (and divisions) often exist between groups. Overlaps may provide early focus for analysis and suggest ways to affect more than one group simultaneously. In some cases, groups may have radically different ideologies but are (or can be) united by a single characteristic. Commanders understand the intricacies of "who does what to whom." Such understanding furthers identifying the urban society's sources of power, influence (both formal and informal), and decisive points that hold the keys to controlling or protecting this potential COG. (See also the discussion of competing power structures in Chapter 3.) Commanders have expert, detailed, and current knowledge and information to avoid developing simple formulas of social interaction that may actively mislead and add to a flawed course of action.

LEADERSHIP AND ORGANIZATION

2-50. Commanders also understand how authority and responsibility is held or shared within and between each of the identified groups. For groups to exert meaningful influence, leadership provides vision, direction, and organized coherence. This leadership can be a function of personality as well as organization. Some groups depend on a charismatic leader to provide cohesion. Others de-emphasize individual leadership and provide redundancy and replacement in decisionmaking. Others combine elements of both these types of leadership and organization. Based solely on personality, a leader may centralize power or, while still being in ultimate control, decentralize decisionmaking and execution to subordinates. In contrast, a single person may head a group while a ruling council actually makes and executes policy. Groups centered on one leader (which may or may not be the officially designated leader) can often produce decisions and initiate actions rapidly but are vulnerable to disruptions if key personalities are removed or co-opted. Groups with shared or redundant leadership take longer to make decisions yet are more resistant to change and outside influence.

INTERESTS AND ACTIONS

> *Me and Somalia against the world, me and my clan against Somalia, me and my family against the clan, me and my brother against my family, me against my brother.*
>
> Somali Proverb

2-51. Identifying and analyzing groups also helps commanders focus on specific segments of the urban society to determine their *beliefs*, *needs*, and *agendas*. It also helps commanders determine how those interests motivate groups to future *action* (or inaction)—previous patterns of activity are critical in this regard. This analysis seeks to determine why groups (and their leaders) act as they do. Commanders consider political, economic, cultural, and religious factors in this analysis. These factors affect all groups to some extent and often provide the basis for their beliefs, needs (actual or perceived), and subsequent behavior. Size and location considerations also apply to each group to help determine to what extent its beliefs or ideologies, needs, and actions may impact the urban operation. However, size and proximity may not accurately indicate actual or potential capabilities. Individuals, small groups, and groups located some distance from the actual conduct of the urban operation may be able to influence large portions of the population. These individuals or groups may have a capability disproportionate to their size and proximity—especially against objectives that are not terrain oriented (as in the case of many stability operations).

INTERACTION, INFLUENCE, OR CONTROL

2-52. As shown above, commanders cultivate an understanding of a group's—

Figure 2-10. UO–Society Cycle of Effects

important. In this instance, the urban commander may need to adjust his IO (to include PSYOP), public affairs (PA) activities, and CMO to counter this propaganda while diverting other combat power to control the populace.

URBAN INFRASTRUCTURE

2-55. Urban infrastructures are those systems that support urban inhabitants and their economy. They link the physical terrain to the urban society. Destroying, controlling, or protecting vital parts of the infrastructure can isolate a threat from potential sources of support. A threat force operating in an urban area may rely on the area's water, electricity, and sources of bulk fuel to support his forces. This is true particularly when his bases or facilities are physically located in or near the area. Isolating this threat from these sources may require him to generate his own electricity and transport his own water and fuel from outside the urban area. To transport supplies, the threat may rely on roads, airfields, sea- or river lanes, and rail lines. Controlling these critical transportation nodes may prevent the threat from resupplying his forces. The control of key radio, television, and newspaper facilities may isolate him from the urban populace (another potential source of support).

INTERDEPENDENCE

2-56. Commanders understand that destroying or disrupting any portion of the urban infrastructure can have a cascading effect (either intentional or unintentional) on the other elements of the infrastructure. Yet, they may be able to gain an operational advantage while minimizing unwanted effects. Commanders can seize or secure an essential facility or structure by using precision munitions, electronic disruption of communications, or SOF and conventional ground forces. To gain this advantage, commanders will rely more on the expertise of engineer and civil affairs units, local urban engineers and planners, and others with infrastructure-specific expertise. After understanding the technical aspects of the area's systems, they can develop the best course of action.

SEPARATE PARTS OF A WHOLE

2-57. Hundreds of systems may exist. Each system has a critical role in the smooth functioning of the urban area. Simple or complex, all systems fit into five broad categories (see Figure 2-11 on page 2-20). Commanders analyze key facilities in each category and determine their role and importance throughout all phases of the urban operation. This analysis considers each infrastructure system individually and in relation to others to determine an appropriate course of action toward it.

STRUCTURES AND PEOPLE

2-58. As depicted in Figure 2-1, each element of the infrastructure consists of both a terrain (physical) and human component. For example, the physical component of the electrical segment of the energy infrastructure consists of power stations, substations, a distribution network of lines and wires, and necessary vehicles and repair supplies and equipment. The human component of this same segment consists of the supervisors, engineers, linemen, and electricians who operate the system. Commanders understand and recognize the physical and human components in their assessments.

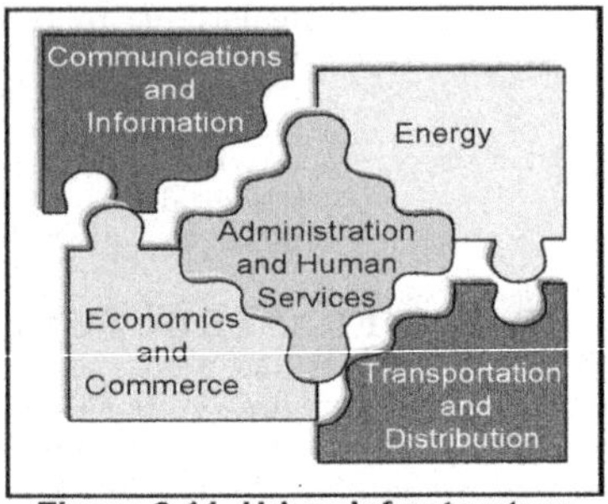

Figure 2-11. Urban Infrastructure

IMPACT ON FUTURE OPERATIONS

2-59. Destroying or incapacitating of any of these elements may impact future operations and inhabitants of the urban area. Destroying urban infrastructure during initial phases of an operation may require commanders to assume responsibility for repair, maintenance and clean up, and operation of those same facilities later. Although exceptions will exist, commanders cannot destroy or significantly damage the infrastructure of a foreign urban center during operations and expect the population to remain friendly to US or allied forces. Still, support from the urban society (albeit of increased importance in UO) is only one factor that commanders weigh while developing appropriate courses of action.

RESOURCE INTENSIVE

2-60. Requirements to protect, restore, or maintain critical infrastructure may divert substantial amounts of resources and manpower needed elsewhere and place additional constraints on subordinate commanders. Civilian infrastructure is often more difficult to secure and defend than military infrastructure. The potentially large and sprawling nature of many systems (such as water, power, transportation, communications, and government) make their protection a challenge. Yet, the infrastructure of an urban area can provide commanders with essential logistics and combat service support. Therefore, the initial expenditure of time and other resources may be necessary to support concurrent or future operations. Legal considerations, however, may affect using the infrastructure and acquiring the urban area's goods and services. Commanders, their staffs, and subordinates (often down to the individual soldier) know their limits concerning Army authority to commandeer civilian supplies or equipment to facilitate mission accomplishment (see the legal support discussion in Chapter 9). In stability operations and support operations, the safeguard or restoration of critical urban infrastructure for military or civilian use may be a decisive point in the overall operation.

COMMUNICATIONS AND INFORMATION

2-61. This system is comprised of the facilities and means to transmit information from place to place. It includes—

ports, and major road networks; ready access to power sources and telecommunications facilities; as well as access to existing local media structures. Hence, media presence may be pervasive and IO even more critical to success in UO than operations in many other environments.

2-65. **A Complex Relationship.** A complex relationship exists among information, the public, and policy formulation. Although the *degree and manner* in which public opinion shapes government policy are uncertain, negative visual images of military operations presented by the media can change political objectives and, subsequently, military objectives. As important, media reporting can influence civilian activity in an urban AO to either the advantage or disadvantage of the commander.

2-66. **Induce Cooperation Through Credibility.** Commanders do not control the media; however, they monitor the flow of information that the news media receives and subsequently reports. Consequently, commanders plan and execute PA operations that will induce cooperation between the media and Army forces. Successful relations between urban Army forces and the news media evolve from regular interaction based on credibility and trust. More information is usually better than less, except when the release of such information may jeopardize security and the success of the operations and threaten the safety of soldiers. However, commanders cannot simply withhold information to protect the command from embarrassment. They consider media interests as part of the normal planning process and work to ensure that information presented to the news media is accurate, timely, and consistent with operations security. Since the media will likely arrive in the urban area before the conduct of operations, early deployment of PA assets may be critical. Commanders synchronize PA activities with CMO and PSYOP. Such action eliminates duplicated effort and ensures a unity of purpose consistent with the IO concept of support (see Chapter 4 for more details involving IO and PA during UO).

2-67. Failure to provide sufficient information can hamper a commander's ability to conduct the mission. Poor relationships with the media can result in inaccurate and even biased reporting. Such reporting can cause a public reaction that influences the ability to achieve operational objectives. During the Russian 1994-95 battle against Chechen separatists in Grozny, for example, the Russian military refused to communicate with reporters. The media reported primarily from the perspective of the Chechen rebels. This encouraged both local and international support for the rebels. It also allowed the Chechens, who lacked sophisticated information systems, to use the media to broadcast operational guidance to their forces. (During their second Chechnya campaign of 1999-2000, Russia learned this lesson well and the Russian view of the war dominated domestic public opinion.) On the other hand, successfully engaging the media can serve as a force multiplier. The Army's open and responsive interaction with the media during peacekeeping operations in Bosnian urban areas helped to explain the challenges and successes of Army forces in the Balkans to the public. This helped maintain domestic, international, and local political support for NATO operations and, with a successful command information program, helped maintain soldiers' morale.

TRANSPORTATION AND DISTRIBUTION

2-68. This element of the infrastructure consists of—

2-72. Sources of energy may be tens or hundreds of miles away from the urban area itself. Therefore, commanders may exert control without applying combat power directly to the urban area itself by controlling or destroying the source (power generation or refinement plant) or the method of distribution (pipe- or power lines). With electrical energy that cannot be stored in any sizable amount, the latter may be the best means as most major urban areas receive this energy from more than one source in a network of power grids. However, control may be as simple as securing a power station or plant and turning off switches or removing a vital component that could later be restored. On the other hand, lengthy pipe- and power lines may compound security and protection of this element of the infrastructure.

2-73. The number of nations that have invested in nuclear power and nuclear research is increasing. With this increase, the potential for Army forces to operate in urban areas that include (or are near) these facilities also increases. Damage to one of these facilities and potential radiation hazards will present special challenges to commanders of a major operation. To safeguard friendly forces and civilians, commanders will need to employ a blend of peacetime and tactical nuclear contamination avoidance principles (see FM 3-14).

ECONOMICS AND COMMERCE

2-74. This system encompasses—

Although his defense of the capital failed, Choltitz succeeded in harboring his available resources, reducing the effectiveness of the resistance organizations fighting his soldiers, and maintaining withdrawal routes for units north of the Seine. The German commander's analysis in support of these efforts was effective in part because of his insightful (1) identification of critical points that included elements of terrain, citizenry, and infrastructure; (2) understanding of the relationships between these parts; and (3) use of an asymmetric approach to address his lack of sufficient force to otherwise handle the densities that challenged him.

Chapter 3

Urban Threat

. . . [T]he United States could be forced to intervene in unexpected crises against opponents with a wide range of capabilities. Moreover, these interventions may take place in distant regions where urban environments, other complex terrain, and varied climatic conditions present major operational challenges.

Quadrennial Defense Review Report, 30 September 2001

As the strategic environment has become less stable, more uncertain, and more dangerous, Army forces are trained and ready to address urban threats. These threats range from regional conventional military forces, paramilitary forces, guerrillas, and insurgents to terrorists, criminal groups, and angry crowds. Although uncertain about events, Army forces can be clear about trends. Increasingly, the Army will face threats that severely differ in doctrine, organization, and equipment, yet can fully interact with the three other components of the urban battlefield—terrain, society, and infrastructure. In stability operations and support operations, commanders broaden their concept of the threat to include natural disasters, hunger and starvation, and rampant disease. Further, commanders plan to contend with many passive urban threats, such as psychological illnesses and toxic industrial materials (TIM). These threats may be found in isolation, but most likely commanders will encounter them in various combinations. Moreover, each new threat will pose a different combination and likely have new capabilities that previous opponents lacked.

CONTENTS

ASYMMETRY

3-1. An emphasis on asymmetric means to offset United States (US) military capability has emerged as a significant trend among potential threats and become an integral part of the threat principles and tactics discussed below. Asymmetry results when one opponent has dissimilar capabilities—values, organization, training, or equipment—that the other cannot counter. It is not a new concept. It naturally evolves from a sound mission, enemy, terrain and weather, troops and support available, time available, civil considerations (METT-TC) analysis by an intelligent, freethinking, and adaptive threat. These asymmetric approaches will include the most advanced, commercially-available technology innovatively applied and mixed with crude, simple, and unsophisticated weapons, tactics, techniques, and procedures.

WEAPONS OF MASS DESTRUCTION

3-2. A chief asymmetric means of engaging the national power of the US is to employ weapons of mass destruction (WMD) against the US or its allies. These weapons can be used against military forces by military forces and include high-yield explosives as well as nuclear, biological, and chemical weapons. Operations in urban areas may require concentrating forces and may create a lucrative target for a threat that possesses fewer numbers and less equipment.

3-3. A threat's WMD use will adversely affect the Army's abilities to conduct urban operations (UO) to various degrees. For example, the intervening structures and the effects of urban microclimates complicate the ability to detect and identify radiological, chemical, or biological attacks from a standoff distance. Also, the individual soldier's ability to recognize his leaders, understand oral and visual commands, and operate increasingly sophisticated equipment is difficult when wearing protective clothing and equipment—particularly if his training proficiency is low. Despite the increased challenges and complexity, Army forces have the training and equipment necessary to respond to such an attack compared to most armies around the world, but certainly when compared to the civilian sector.

3-4. Although initial casualties could be high, the public can accept military casualties before those of civilians. Therefore, threats may gain an initial tactical advantage but would achieve less asymmetric benefit by directly attacking Army forces. They may attempt to achieve an extraordinary asymmetric strategic advantage by employing WMD against US or allied civilian populations. In doing so, threats hope to use political sensitivity to high civilian casualties to reduce popular support for the US or its allies. The chance of these attacks occurring in an urban area increases because—

THREAT OPERATIONAL PRINCIPLES

3-5. The threat may apply several key operational principles to oppose Army forces operating in an urban environment (see Figure 3-1). These principles focus more on how a threat might fight in an urban area rather than specifically whom the threat might be or in what region of the world the conflict might occur. They are more effective in an urban environment due to—

NEUTRALIZE TECHNOLOGY OVERMATCH

3-8. Threats will always strive to force engagements at a time and place most advantageous to them. They may locate military forces and vital military capabilities in urban areas to achieve sanctuary from the effects of Army capabilities and make Army forces and systems more vulnerable to less-sophisticated weapons.

3-9. The clutter of the physical structures, electromagnetic radiation, and population diminishes Army capabilities. This clutter makes it difficult for Army forces to acquire and effectively engage targets at long ranges. In urban areas, the terrain often allows a threat to operate in closer proximity to friendly forces. Therefore, the threat may "hug" friendly forces to avoid the effects of high-firepower standoff weapon systems and degrade their ability to gain or maintain a thorough common operational picture. Additionally, this threat tactic attempts to inhibit friendly commanders from employing some weapon systems and munitions for fear of fratricide.

CONTROL THE TEMPO

3-10. Threats will try to achieve a decisive advantage by setting and controlling the tempo necessary to achieve their objectives. To prevent the Army's entry into theater, threats may try to create a high operational tempo to take advantage of the inherent weaknesses in power projection operations outlined earlier. As other efforts deny entry, threats may seize the initiative, achieve surprise, and exploit the tempo differential by attacking with heavy conventional forces potentially possessing greater firepower and more rapid ground mobility than the Army's initial-entry forces.

3-11. If they cannot deny entry or end the conflict quickly, threats may use any preparations made in the initial high-tempo period to prolong the event, aiming to degrade US or allied commitment. The complex nature of the urban environment slows operations conducted in and around these areas. Threats may maximize this characteristic by fighting for key urban complexes and infrastructure, forcing friendly forces to operate within these areas. If Army operations focus on one or more urban areas, the overall campaign slows. However, even when UO make up only one component of a much larger campaign, they may consume valuable resources needed for other operations and delay the entire campaign.

Tempo

The battle for Aachen, Germany, in the fall of 1944, developed during the US First Army's offensive to breach the Westwall fortifications. Aachen, the ancient capital of Charlemagne, had symbolic political and psychological significance for the Germans and Americans. Furthermore, it was the first city on German soil to face an assault by the Allies. Consequently, the symbolic importance of this first major battle in Germany ensured bitter resistance against American attackers. The Germans surrendered only after the city was destroyed. Expected to take a few days, instead, the battle took weeks. Although the Army had achieved a clear tactical victory, the German defense of Aachen cost the First Army valuable time and resources, and delayed the planned attack to the Rhine River.

CHANGE THE NATURE OF THE CONFLICT

3-12. Threats may attempt to change the fundamental nature of the urban conflict to exploit ambiguous or tenuous political-military objectives. Many nations gain and maintain domestic popular support to use their armies for political objectives. The threat may attempt to change the nature of the conflict by modifying its strategy and tactics, the environment, or any combination, ultimately hoping to reduce friendly popular support. For example, introducing an urban terrorist threat to US civilians or soldiers not directly engaged in operations changes the nature of the conflict. This type of threat may not have been an initial consideration, and this change may reverse public support for the operation. Another example, growing US coalition combat power may cause the threat to switch from open maneuver warfare to UO to avoid decisive combat with superior forces and achieve a stalemate. Originally expecting a quick solution or victory, the political leadership may now envision a longer deployment with less chance of lasting success.

CAUSE POLITICALLY UNACCEPTABLE CASUALTIES

3-13. Threat forces may gain an advantage against superior friendly forces by capitalizing on a perceived weakness of many Western nations: the inability to endure continuous losses or casualties for other than vital national interests or losses for which they are psychologically unprepared. A secondary US interest may equate to national survival on the part of a threat. Therefore, the threat (particularly with fanatical leadership) may willingly sacrifice excessive amounts of money, equipment, and people (soldiers and civilians) to achieve victory. Threats may attempt to weaken US resolve and national will to sustain the deployment or conflict by inflicting highly visible, embarrassing, and if possible, large losses on Army forces, even at the cost of similar losses to themselves. Many threat forces will use UO to inflict mass casualties and destroy premier Army weapon and information systems. The physical characteristics of the urban environment support these ambush techniques. Light infantry or insurgents with readily obtainable, hand-held antiarmor weapons can effectively attack armored vehicles and helicopters, no matter how sophisticated, in an urban area.

ALLOW NO SANCTUARY

3-14. Threats will attempt to deny Army forces safe haven anytime and anywhere. Terrorism may be one of the tactics used to deny sanctuary to Army forces. They will attack Army forces anywhere, particularly while operating in urban areas where the fear from being attacked from any quarter is often greater. Threats may be or employ state-sponsored or independent terrorists, well equipped and motivated to accomplish their assigned missions.

3-15. Military buildings, facilities, and installations in urban areas are particularly vulnerable to high-yield explosive munitions as well as other clever means to create large explosions. The close-in nature of urban areas, large populations, and high volume of vehicle traffic provide a good environment for target reconnaissance, explosives positioning (conventional or high-yield), and cover for an attack. These attacks will likely be preceded by extensive, careful reconnaissance, necessitating a solid friendly counterterrorism and counterintelligence effort.

CONDUCT DISPERSED AND DECENTRALIZED OPERATIONS

3-16. To a certain extent, dispersed and decentralized operations are an integral part of all threat principles. However, this concept warrants separate emphasis as a principle since threat forces will likely place great significance on it on future urban battlefields. Both dispersed and decentralized approaches seek to reduce threat vulnerabilities to air power and precision-guided munitions (PGM) while increasing their agility, flexibility, and overall maneuverability in an urban environment.

3-17. Urban terrain tends to fragment and separate forces that operate in it. Threat forces recognize this characteristic, accept it, and make it work to their advantage. They conduct operations from dispersed urban locations to reduce their vulnerability to friendly decisive operations and massed firepower. Although separated, threat forces will attempt to retain the ability to assemble and mass quickly so to strike as opportunities present themselves. Once threat forces complete the operation, they will return to separate locations to avoid potential counterattack. The fluidity and seemingly disjointed appearance of these threat UO will challenge friendly efforts to conduct templating and pattern analysis. Ambushes (air and ground) will be used to deny friendly ground and air reconnaissance of their dispersed locations.

3-18. Dispersed operations normally depend on good command and control to achieve synchronization and massed effects. Threat forces also understand the debilitating effects of the urban terrain on communications and the execution of operations. When they cannot mass their forces or effects, they will depend on decentralized operations to achieve their objectives. They will operate autonomously, guided only by a higher authority's purpose and intent. These operations make them even less vulnerable to massed attacks and PGM as smaller threat forces do not present an objective or target that will allow friendly decisive operations. Again, pattern analysis and templating will be extremely difficult. Using this principle often prolongs the conflict but is central to implementing the other threat principles.

URBAN THREAT TACTICS

3-19. Urban areas provide a casualty-producing and stress-inducing environment ideally suited for using specific urban threat tactics. Moreover, urban areas provide threats with an unmatched degree of cover and concealment from friendly information and firepower systems. While active urban threats may vary widely, many techniques will be common to all. Figure 3-2 outlines a set of threat tactics available to potential threats opposing mission accomplishment in urban areas. *Army forces may use many of the threat tactics, except those that violate the law, ethics, and morals, to defeat urban threats.* Moreover, using asymmetric means is not the sole domain of the threat. Army commanders can also leverage capabilities, create conditions, and plan operations to develop asymmetric advantages to accomplish the mission.

USE THE POPULATION TO ADVANTAGE

3-20. Many urban areas may be too large to evacuate completely (if at all). Even if desirable, a military force may have no place to safeguard and secure the inhabitants. Therefore, future UO may see large segment of the populace remain. Offensive and defensive operations may be constrained not only by the terrain and by the presence of many civilians. Army forces involved in urban stability operations and support operations will certainly conduct missions in and amongst the residents. These residents may restrict operations and, when gathered in large numbers, may (even without initial hostile intent) present a critical force protection issue for the commander.

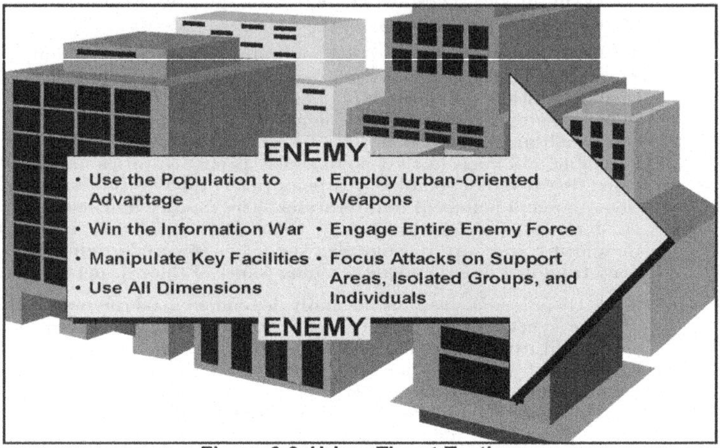

Figure 3-2. Urban Threat Tactics

Use as Key Terrain and Concealment

> *Chechen fighters sometimes disguised themselves as Red Cross workers, donning the identifying armbands. They also passed themselves off as civilians and offered to guide Russian forces through the city, instead leading them into ambushes.*
>
> Olga Oliker
> *Russia's Chechen Wars 1994-2000*

3-21. From the threat standpoint, the populace is similar to key terrain: the side that manages it best has an advantage. Threat forces may gain this advantage by using civilians as camouflage, concealment, and a means of deception. Guerrilla and terrorist elements may look no different from any other member of the community. Many foreign conventional and paramilitary troops often have a "civilian" look. Western military forces originally adopted the clean-shaven and close-cut hair standards to combat disease and infection, but future opponents may not adhere to those standards. They may adopt grooming standards, civilian-looking clothing, and other "nonmilitary" characteristics to make themselves indistinguishable from the civilians.

Identifying Soldiers from Civilians

During Russia's 1994-95 conflict with Chechnya, Russian forces had difficulty identifying Chechen guerrilla forces from Grozny's noncombatant population. Because their appearance was identical to that of the urban populace, Chechen soldiers could freely walk around the city, suddenly disappear, and then abruptly reappear firing their weapons from basements, windows, or dark alleyways. To distinguish fighters from peaceful city dwellers, Russian forces began looking at men's shoulders to see if they were bruised (from firing weapons) and their forearms to see if there was burned hair or flesh (from the extraction of cartridges). They closely examined their clothing and smelled for gunpowder residue. To identify a Chechen artilleryman, Russian soldiers checked for glossy spots left by artillery and mortar rounds on the bends and cuffs of sleeves. They also turned pockets inside out to check for a shiny, silvery-leaden hue indicating the former presence of small arms ammunition. Russian forces also recognized a grenade launcher operator or mortar man from fibers and crumpled pieces of gun cotton on their clothing. US Army commanders may need to develop similar, imaginative means to identify the threat.

Gain Cover, Protection, and Increased Mobility

3-22. Threat forces may attempt to gain cover by using the urban inhabitants as human shields. With this increase in protection, they simultaneously increase their mobility. They recognize the Army's focus on developing and applying rules of engagement (ROE). They will take advantage of the restraining effects of international law and the Army ethical values to enhance their mobility in proximity to friendly positions. Knowing the Army's reluctance to cause noncombatant casualties and collateral damage, threats may operate in areas containing civilians and essential facilities to restrict the Army's use of massed or nonprecision firepower. They may also employ "rent-a-crowds"—civilians paid to demonstrate against military forces—armed only with sticks, stones, and Molotov cocktails (a potential asymmetric challenge).

Make Moral Responsibilities a Weakness

3-23. Depending on their successes, threats may use these tactics and skillful information operations that attack national will and coalition sensitivities in an attempt to force the Army to establish more restrictive ROE. Threat forces may also take advantage of the Army's moral responsibilities. By herding refugees into friendly controlled areas, threat forces try to make the civilians a burden on the Army's logistic and security resources. Threat forces, on the other hand, may not abide by international agreements, such as the Geneva conventions. They may not take prisoners unless they can be ransomed or made part of a local prisoner exchange. They may even execute friendly prisoners in front of the media to show their "strength" and, more importantly, to cause friendly forces to overreact and lose their legitimacy. Threat forces can then use such an overreaction to unite others with their cause.

Acquire Intelligence and Logistic Support

3-24. Indigenous threat forces can normally use the local population for intelligence and logistic support far more effectively than can an alien army.

Threat forces may manipulate local hires serving among US soldiers, such as those contracted by the Army for base operation purposes or translator duties. In addition, refugees moving through friendly controlled sectors may provide the threat with information on friendly dispositions, readiness, and intent. Even friendly residents may become unwitting or unwilling informants, providing an enemy or a hostile with vital information on friendly activities, dispositions, and capabilities. However, a threat employing particularly cruel, abusive, or repressive measures may easily turn certain groups in the urban area against them, even when they share a common history, culture, and ethnicity with the civilians. This is more likely in those areas with high population densities.

3-25. Threat forces may also seek to use some nongovernmental organizations (NGOs). They may try to obtain relief supplies either through the organizations' legitimate relief operations or as a target for theft. Some organizations may even be fronts for weapons, food, ammunition, money, and fighters. For example, during Russia's second conflict in Chechnya (1999-2000), documents purportedly found in Grozny by the Russians listed nations such as Sudan, Nigeria, Niger, and Ivory Coast as sending fighters to Chechnya under the guise of the International Islamic Relief Organization. (Chechen fighters also disguised themselves as Red Cross workers.) This deception increases the need for strict security and force protection measures, close coordination with NGOs operating in urban areas, and closer monitoring of suspect organizations' activities by civil affairs personnel.

WIN THE INFORMATION WAR

3-26. Threat forces will try to win the information war as much as they will directly oppose Army UO. Threat urban campaigns need not be tactical military successes. They need only to weaken legitimacy and make the opposition's campaign appear unpalatable to domestic and world support. As a critical part of their overall information operations, threats will use the ever-present media to tell their story. Portable video cameras, commercial radios, and cellular telephones, available and easily concealed, will be as important to many threat actors as weapons and ammunition. Internet access, already firmly established in many urban areas, provides the means to easily disseminate threat propaganda, misinformation, and disinformation through web sites and electronic mail. Hackers, covered and concealed in the interior spaces of the urban area, may gain access to US sites to manipulate information to the threat's advantage.

Information and the Media

The media coverage of the urban battle for Hue, South Vietnam, although only one of hundreds of different attacks of the Tet Offensive, affected the will of both the American people and their political leadership. On January 31, 1968, two North Vietnamese Army (NVA)/Vietcong (VC) regiments and two sapper battalions, moving rapidly and with the element of surprise, attacked and seized part of the walled city (Citadel) of Hue. It was the third largest city in South Vietnam, the former capital of a united Vietnam, the capital of Thua Thien province, and a spiritual and cultural center. Initially intending to hold the city for seven days, the NVA/VC retained portions of the city for approximately three weeks against determined US and South Vietnamese attempts to retake it.

Hue marked a revolution in the coverage of war by modern media. It was the first time Americans could sit at home and watch an ongoing battle on the evening news. One of the most intense and savage battles of the Vietnam conflict, it was televised every evening for almost a month. Although the battle for Hue was a tactical victory for the US, the North Vietnamese clearly achieved strategic success by searing the American consciousness with the high costs of urban warfare. Had US leaders made winning the information war a central part of the overall campaign plan—for example, exposing the American people to the NVA's brutality by publicizing the civilian executions in Hue—civilian support for the war may have been bolstered and a different outcome achieved. See Chapter 6 for a more detailed account of the battle for Hue.

MANIPULATE KEY FACILITIES

3-27. Threat forces will attempt to identify and quickly seize control of critical components of the urban area to help shape the battlespace. Urban telephone exchanges, for example, provide simple and reliable communications that can be easily secured with off-the-shelf technologies. Sewage treatment plants and flood control machinery can be used to implement WMD strategies or to make sections of the urban area uninhabitable. Media stations significantly improve the information operations abilities of the controlling force. Power generation and transmission sites provide means to control significant aspects of civilian society over a large area.

USE ALL DIMENSIONS

3-28. Threats will think and operate throughout the depth, breadth, and height (including supersurface and subsurface areas) of the urban environment. Conventional lateral boundaries will often not apply as threat forces control some stories of the same building while friendly forces control others.

3-29. Intrasurface areas and roofs provide urban threats with excellent observation points and battle positions above the maximum elevation of many weapons. Shots from upper floors strike armored vehicles in vulnerable points. Basements and other subsurface areas also provide firing points below many weapons' minimum depressions and strike at another weakness in most armor. Sewers and subways may provide covered and concealed access throughout the area of operations.

EMPLOY URBAN-ORIENTED WEAPONS

3-30. Whether purpose-built or adapted, many weapons are more useful in an urban environment while others may have significant disadvantages. Urban threat weapons are much like the nature of urbanization and the urban environment: inventive and varied. Many threats will integrate widely available off-the-shelf technologies into their weapon systems and armed forces. However, sniper rifles and small, man-portable, fire-and-forget weapons and demolitions and other improvised explosive devices will likely dominate the urban environment. Figure 3-3 lists examples of threat weapons favored in UO.

ENGAGE ENTIRE ENEMY FORCE

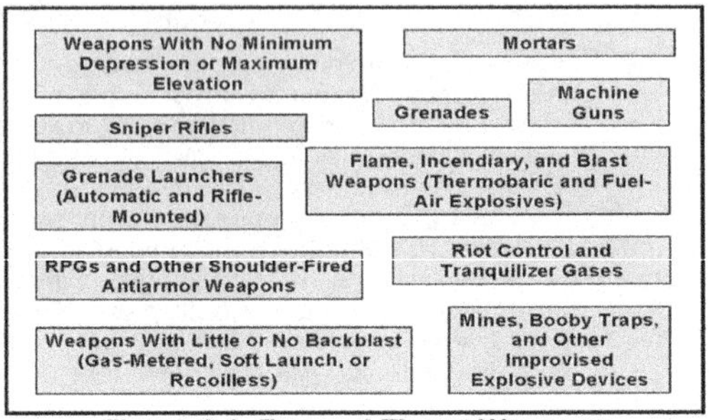

Figure 3-3. Favored Threat Weapons

3-31. Threats may attempt to keep all or significant portions of Army forces engaged in continuous operations to increase their susceptibility to stress-induced illnesses. UO, by their nature, produce an inordinate number of combat-stress casualties. Continuous operations exacerbate this problem. Threat forces that employ this tactic will often maintain a large reserve to minimize the psychological impacts on their own forces.

3-32. To accomplish this, threat UO will likely involve decentralized maneuver, precision fires, and simultaneous operations involving unconventional and special purpose forces. Threat forces will take advantage of any exposed weakness and engage in battles as opportunities present themselves.

FOCUS ATTACKS ON SUPPORT AREAS, ISOLATED GROUPS, AND INDIVIDUALS

3-33. To supplement the previous tactic, threat forces will seek to target support areas, small groups, leaders and their headquarters, and individual soldiers. Their focus on resupply, casualty evacuation, and other sustainment activities, coupled with the compartmented terrain, navigational challenges, and multiple three-dimensional avenues of approach often makes these locations and soldiers more susceptible to surprise raids and ambushes. Attacks on these areas and groups are conducted to erode the Army's ability to sustain UO, to inflict maximum casualties, and to induce psychological stress. These attacks can be mitigated by careful, regular evaluation of choke points and other restrictive terrain, regular awareness training for units and individuals operating in or transiting through potential incident-prone areas, and thorough after-action analysis of incidents.

NEGATIVE EFFECTS OF URBANIZATION

3-34. Many urban areas are the engines for increased industrialism and economic growth as an expanding population provides the labor for manufacturing and service needs. However, rapid and inadequately planned growth can result in undesirable consequences. Uncontrolled urbanization may result in an infrastructure and economic base unable to support the growing population. A large transient, ill-housed, and idle population in a close geographic space may produce strife. Classes, cultures, ethnic groups, and races that might otherwise peacefully coexist can clash under the stress of survival. Uncontrolled urban growth has resulted in the negative effects listed in Figure 3-4 on page 3-12. *In many urban stability operations and support operations, these may be the primary "threats" to mission accomplishment.*

3-35. Not all urban areas prevail as inherently unstable or hotbeds for unrest. Urban growth due to migration may remove sources of conflict, or it may provide the catalyst for violence. Commanders recognize the possible effects of uncontrolled urbanization. During their intelligence preparation of the battlefield (IPB), they determine if these conditions exist. Throughout mission analysis and the development of courses of action, commanders consider the impact (if any) on their operations. At the same time, they recognize that UO may *create* similar problems that may affect the current operation as well as the overall campaign.

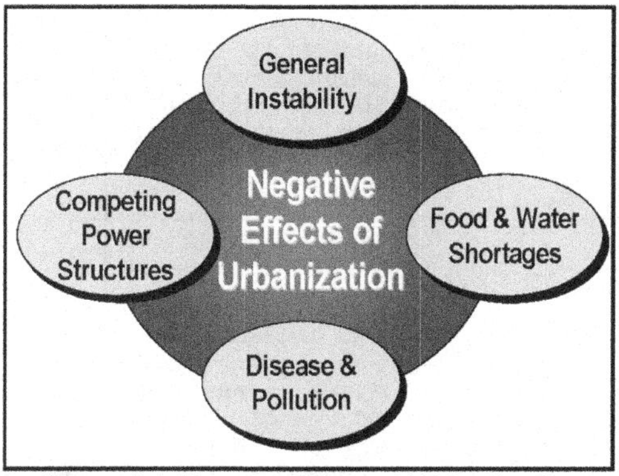

Figure 3-4. Negative Effects of Urbanization

GENERAL INSTABILITY

3-36. Urbanization can enhance stability by generating industrialization and economic growth resulting in more jobs, a higher overall standard of living, and an educated, relatively satisfied populace. However, the population dynamics associated with urbanization can also have an opposite, destabilizing effect. Radical population growth may create overcrowding and generate or aggravate resource and quality of life issues. Intense and destructive competition for employment, housing, and social status may develop in this climate of economic deprivation. The inability of some governments to handle these problems—

already struggling institutions, hasten conflict, and lead to overall instability. Commanders understand that UO, depending on the operation, may either cause massive population movement out of or into urban areas.

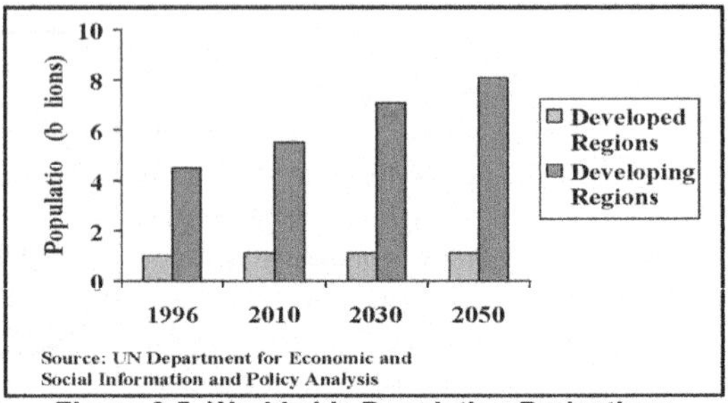

Figure 3-5. Worldwide Population Projections

3-38. Urban areas with a large youth population may also help to generate conditions for instability. Rural-to-urban migrants tend to be relatively young. In 1999, Cairo, for example, had more than 40 percent of its population younger than 15 years. Young urban populations generate enormous demands for social resources, primarily education and jobs. Even a strong urban economy may fold under the economic expectations of a tremendous influx of young migrants. Disorder and violence may result as hostiles (many nonstate actors) easily mobilize and manipulate the idle young to act politically and criminally. Urbanization and population growth are more dangerous when they combine to produce a cohort of young urban dwellers separated from traditional social controls, such as village elders and clan leaders.

3-39. Ethnic, religious, and other social issues may become the vents for anger and frustration produced by the high tension of urban life. Major acts of violence and destruction, such as occurred in 1992 in India, can directly threaten a nation's security. Army forces may have to conduct large-scale, stability operations and support operations to promote peace and protect national interests. In these cases, all levels of command will be particularly concerned with maintaining impartiality and perceived legitimacy.

Cultural and Religious Instability

The 1992 bombing of the Babri Masid Mosque in Ayodya, India, enflamed an already intense cultural and religious rivalry between Hindus and Muslims and led to rioting throughout many Indian urban areas. Of the 1,500 who died in conflicts and riots, almost 95 percent died in urban areas. The violence struck Ahmedabab and Bombay most seriously, with acts of murder, gang rapes, and arson occurring months after the destruction of the mosque.

FOOD AND WATER SHORTAGES

3-40. Rapid urbanization, primarily in developing nations, may lead to severe food shortages that could influence Army forces (or lead to their use). Such shortages may cause instability, massive migration, revolts, or increased support of armed opposition groups. Armed factions may target NGOs that supply aid as a means of furthering dissatisfaction among the populace. In effect, food may become a weapon. Deployed troops may need to provide or support humanitarian food aid networks to keep the humanitarian situation from escalating.

3-41. Normally, commanders should use centralized feeding centers as a last resort. Instead, Army forces should bring the food closest to the population to encourage civilians to stay in their homes. If safe areas or camps are created, they should be designed for use over as short a time as is feasible. The general rule should be to return the urban population to their homes as soon as possible. Army forces conducting domestic support or foreign humanitarian assistance operations that cannot maintain the safe food supplies may find the frustrations and hostility of the local population focused on them.

3-42. Water shortages (and quality) are becoming a serious problem in many regions. Commanders operating in an urban environment need to know the water supply origins and its treatment, purification, distribution, and vulnerabilities. Before beginning operations, commanders must know if they are providing water for the noncombatants as well as their own forces. Across the range of operations, controlling and protecting a limited water supply is, or may become, an essential operational consideration during UO.

Food and Water Shortages

Countries as varied as Indonesia and Algeria exported their food surpluses only two generations ago but now import up to two-thirds of their basic staples. This cycle has resulted in many countries, which once exported agricultural products, facing the growing cost of imports to feed their urban populations. Estimates predict that by the 2010, at least 65 countries (including 30 of Africa's 51 countries) may depend completely on food imports. For some countries, it is even worse. Congo (Zaire), once a net food exporter, now faces mass starvation.

Over the last four decades in China, irrigated farmland has tripled and urban populations have quintupled. In Indonesia, urban areas such as Jakarta may use six times more water in 2005 as it did in 1990, and Indonesia currently has limited capability to meet this increased demand.

DISEASE AND POLLUTION

3-43. Urban areas frequently spawn epidemics; therefore, widespread disease may pose a significant threat to Army forces that operate there. In many developing nations, rapid urbanization has occurred without a corresponding upgrade, expansion, or even development of adequate sewage and water systems. Some urban areas have only one toilet for every 750 people or more. In these areas, hundreds of thousands live much as they would in poor villages, yet so confined as to ensure high transmission rates for airborne, waterborne, sexually transmitted, and contact-transmitted microbes.

3-44. In urban areas lacking adequate trash and waste management infrastructure, insect-spread diseases proliferate. Mosquitoes that breed in polluted water, open water tanks, and irrigated urban gardens carry malaria and dengue fever—the leading causes of sickness and death from infectious disease in Latin America and Africa. The problem compounds with growing numbers of bacteria resistant to various antibiotics, a shortage of trained medical personnel, inadequate or insufficient medical facilities and supplies, and unclean agricultural and food-processing practices.

3-45. Pollution also creates critical health problems in developing areas and a potential health risk for intervening Army forces. Urban areas in China have recorded five to ten times the levels of sulfur dioxide found in the air of urban areas in the developed world. In parts of Poland, toxic waste has so polluted the land and water that ten percent of the babies have birth defects. Pollution may cause immediate health problems but more often, the insidious effects appear months or years after exposure. As discussed earlier, UO may contribute, either intentionally or unintentionally, to an increase in pollution. Destruction of industrial complexes that use, produce, and store hazardous material may produce toxic gas and smoke pollutants that contributed to significant health concerns to exposed soldiers.

3-46. Commanders initiate combat health support (CHS) planning early, including analysis of the medical threat and other critical medical information requirements during the IPB process. A medical surveillance system monitors the daily status of Army personnel throughout the operation. In preparation, all personnel receive a predeployment medical examination. This exam establishes an accurate baseline health status of the force and ensures that Army forces do not introduce new diseases to an urban area, possibly exacerbating the situation. Conversely, soldiers not immune to native viruses or possessing a weakened immune system due to continuous operations and the stress associated with UO may put Army forces at a significant disadvantage. An outbreak of plague during an operation would have an effect similar to a chemical or biological attack. The closer that Army forces operate to civilians (the humanitarian assistance operations conducted in Port-au-Prince, Haiti, and Mogadishu, Somalia, for example), the more probable that these situations may occur. See Chapter 9 for further CHS considerations.

COMPETING POWER STRUCTURES

3-47. Many groups can exist that become strong enough to rival the power of the governing officials and eventually turn the urban area into a system of divergent and competing power structures. These groups can consist of insurgent forces, a merchant class or an economic elite, criminal organizations, or some other significant source of power such as religious organizations, clans, or tribes. In the absence of a legitimate authority, armed factions headed by "warlords" may vie to fill the power void. Sometimes these groups or organizations, normally at odds with each other, may form alliances to achieve specific goals. Commanders recognize, identify, and understand these alternate urban power bases and, if necessary, develop engagement strategies to neutralize or harness them to accomplish the Army mission.

Urban Insurgencies

3-48. As urban migration increases in the developing world, rural guerrillas appear to follow. This transition of insurgencies from rural to urban areas occurs because urban areas offer a rich field of targets for insurgent attacks. People immediately notice any disruption of urban infrastructure, thus having great propaganda value. A concentrated urban population is often more susceptible to propaganda and political organization. Insurgents can easily arrange mass demonstrations using available communications facilities, both overt and covert. Travel is effortless and large urban populations provide cover and concealment. On the whole, urban areas may provide a fertile environment for guerrillas to apply their rural insurgent strategies. However, even with a rural-based insurgency, operations in urban areas offer distinct opportunities to disrupt, discredit, and demoralize the government (see FM 3-05.20 and FM 3-07).

Urban Insurgencies

In Africa, a strategy of capturing urban areas, while trapping government forces within others, has become a common tactic of insurgent forces. Similarly, insurgents in Liberia concentrated their efforts in the capital city of Monrovia while guerrillas in Sierra Leone have battled the government repeatedly for the urban diamond mining hubs. Even Shiite rebels in Afghanistan took their conflict with the government into the heart of Kabul, the capital.

Merchant Class

3-49. Urban areas normally possess a merchant class or an economic elite as part of their social structure. In some urban areas, they may carry more power than the local or central state government. They may isolate themselves physically and socially from the sprawling poor yet wield enormous power over the country's political and economic activities. The degree of economic separation between the merchant class and the poor may be small but still socially or politically significant.

3-50. In a vastly impoverished area where the economy of the urban area is severely disrupted, the merchant class will often continue to operate and function and, as a result, achieve a measure of influence. To continue to operate under acute economic turmoil, they may form alliances in criminal organizations and secure loyalties within the government. Outside resources introduced into a crisis area (such as food, water, fuel, and pharmaceuticals) take on increased value, may replace currency as the medium for exchange, and often become the means to amass and hold wealth. One of the primary ways to obtain wealth may be to steal it.

3-51. In some turbulent situations that lead to the need for stability operations or support operations, commanders may harness the power of the merchant class as a force for peace and stability instead of one that uses crime to achieve economic goals. For example, in a relief situation, instead of competing with the merchant class by distributing food directly to the needy and possibly creating an environment of looting and black marketeering, it may be possible to monetize food. Food assistance from donor governments could be sold to merchants at an attractive price so they have a reliable source of supply. This could, in turn, create a healthy economic system and separate merchants from criminals and gangs.

Criminal Organizations

3-52. Organized criminal groups have grown common in urban areas; have also become an important part of the urban social structure (gangs for example); and can exert considerable influence on governments, people, and military forces conducting UO. Some large criminal organizations relying on international connections often have better resources and equipment than their insurgent counterparts. Their large financial resources, long-reaching connections, and ruthlessness provide them the means to corrupt or intimidate local officials and government institutions. In any operation, but especially support operations, they may violently confront and oppose Army forces during mission execution.

3-53. The tactics of urban criminal groups parallel those of insurgents. They have developed an intuitive cultural understanding of slum neighborhoods and the ability to lure civilians into criminal activities. They have also mastered the management of mobs. They recruit teenagers and young adults in their efforts against rivals and authorities, just as insurgents muster armies from the youth of rural villages. In many developing nations, there exists an alliance between insurgents and organized criminal groups. In these alliances, the insurgents defend the criminals and the criminals fund the insurgents. During many UO, particularly during or following combat, civil disturbances, or large natural disasters, looting (organized or unorganized) may become of critical concern. Therefore, UO may often require a combined law enforcement and military response.

Crime and Criminal Organizations

Crime and poverty plague urban areas such as Rio de Janeiro, Brazil's second largest urban area and would affect military operations conducted in their limits. Rio has some of the nation's highest negative urban indicators: the largest number of slum dwellers (1 million), the highest murder rate (1 of 700 residents per year), and the highest kidnapping rate (4 per week). In 1989, the homicide rate of the urban area was three times higher than New York City's and the rate of urban violence continues to rise. Therefore, law enforcement management may be a critical issue for Army forces operating in urban environments similar to that of Rio de Janeiro.

However, criminal elements or organizations may not always work against Army commanders. They can be co-opted or influenced to serve friendly objectives. For example, during World War II the US Navy worked covertly with the Mafia in New York City to secure the New York harbor from German U-boats believed to be torpedoing ships there. The Mafia controlled most dock activities New York harbor and was perfectly positioned to monitor other subversive waterfront activity. This capability provided needed information to the Navy for its counterintelligence and security tasks. New York civil authorities therefore agreed to permit a Navy-Mafia alliance to operate at the port for the greater good of the country. Although the Mafia was not the preferred ally of the Navy, it had the capability to protect US ships and the interest (patriotism) to help in the war effort. In those circumstances, the temporary alliance worked (see also the civilian threat discussion in Appendix B).

Warlords

3-54. A characteristic of many recent stability operations and support operations has been the deterioration or complete collapse of political authority in the country or urban area in crisis. In some cases, warlords have attempted to fill the power vacuum (see Appendix C for an example). These individuals often have no particular claim to legitimacy. Their power issues from their weapons, not necessarily from their political skills, human services provided, or popular consent (although they have *some* popular support to remain in their relative position of authority). In dealing with these urban warlords during support operations or stability operations, it may appear that there are two options: either ignore them completely or work with them visibly or regularly. Commanders may reduce some of the greater risks involved in these extremes by adopting a middle-ground approach. Nevertheless, the technique chosen must clearly support political and military objectives.

3-55. Refusal to acknowledge warlords may increase the threat to Army forces and NGOs. Their militias may attack Army forces to achieve recognition or simply due to misunderstanding or inherent friction between armed forces. On the other hand, dealing with them may provide legitimacy to the exclusion of other elements of the urban population such as professional groups (for example, doctors or teachers), religious groups, and traditional clan or tribal chiefs—which may have a greater claim to legitimacy and better form the foundation for a reconstituted urban society.

3-56. A compromise between these two extremes may offer the best chance for success. Commanders generally recognize these warlords or they risk incidents; however, this recognition can be kept at staff levels to avoid bestowing any legitimacy on them. Instead, commanders themselves visibly meet the other elements of society that have a more legitimate claim to political, social, or economic leadership. Inevitably, commanders may need to meet with warlords. In those circumstances, clan or tribal elders, and others who represent traditional authority should attend and commanders should ask for, and give deliberate consideration to, their opinions.

Chapter 4

Contemplating Urban Operations

We based all our further calculations on the most unfavorable assumptions: the inevitability of heavy and prolonged fighting in the streets of Berlin, the possibility of German counter-attacks from outside the ring of encirclement from the west and south-west, restoration of the enemy's defence to the west of Berlin and the consequent need to continue the offensive.

General of the Army, S. M. Shtemenko
describing the operational level planning for taking Berlin
The Soviet General Staff at War

A *major operation* is a series of tactical actions (battles, engagements, strikes) conducted by various combat forces of a single or several services, coordinated in time and place, to accomplish operational, and sometimes strategic objectives in an operational area.

In any potential situation and in any area, Army commanders will likely need to assess the relevance and impact of one or more urban areas on their operations. They will also need to determine whether full spectrum urban operations (UO) will be essential to mission accomplishment. UO may be the commander's sole focus or only one of several tasks nested in an even larger operation. Although UO potentially can be conducted as a single battle, engagement, or strike, they will more often be conducted as a major operation requiring joint resources. Such actions result from the increasing sizes of urban areas. Army commanders of a major urban operation then ensure that UO clearly support the operational objectives of the joint force

CONTENTS

commander (JFC), requesting and appropriately integrating critical joint resources. Whether the urban operation is the major operation itself or one of many tasks in a larger operation, Army commanders assess and thoroughly shape the conditions so subordinate tactical commanders can dominate in the complex urban environment.

NECESSITY OF URBAN OPERATIONS

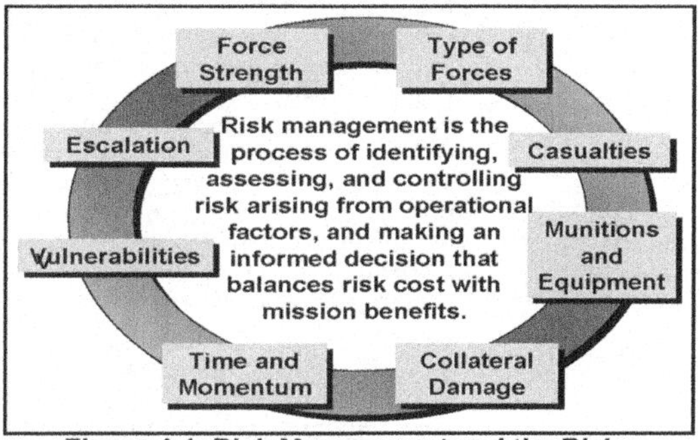

Figure 4-1. Risk Management and the Risks Associated With Urban Operations

4-1. Early in planning, commanders of a major operation address the necessity of conducting operations in urban areas located throughout their areas of operations (AOs). Chapter 1 discussed strategic and operational considerations that compel forces to operate in urban areas. These reasons include the location of the threat force; critical infrastructure or capabilities that are operationally or strategically valuable; the geographic location of an urban area; and the area's political, economic, or cultural significance. Several considerations exist, that may make UO unnecessary, unwarranted, or inefficient. When determining whether to operate in an urban environment, commanders consider the operational (and accidental) risks and balance them with mission benefits. The factors shown in Figure 4-1 highlight some measures to evaluate the risks associated with UO.

FORCE STRENGTH

4-2. When facing prospective UO, commanders consider if they have troops available to conduct the operation properly and with acceptable risk. Under normal circumstances, large urban areas require many forces merely to establish control. New York City police department has over thirty thousand officers simply to conduct peacetime law enforcement. Major UO, particularly those that are opposed, will often require a significant number of forces. If commanders lack sufficient force to conduct effective operations, they may postpone or consider not initiating those operations until they have the necessary strength. Commanders add to their analysis the requirements for troop strength elsewhere in the AO.

TYPE OF FORCES

4-3. Along with force strength, commanders consider the type of forces available. This consideration includes an assessment of their level of training in urban operations. *All UO put a premium on well-trained, dismounted infantry units.* Therefore, Army forces conducting UO should be force tailored to include a large infantry component. In addition, special operations forces (SOF) are invaluable in UO. SOF include psychological operations (PSYOP) and civil affairs (CA) forces. They should always be considered as part of the task organization.

4-4. UO include combined arms to ensure tactical success in combat. Although masses of heavy forces are not normally required, successful UO require all the combined arms capabilities of all Army forces. Even if an urban operation is unlikely to involve offensive and defensive operations, field artillery may be essential to force protection. In urban stability operations and support operations, successful mission accomplishment requires more robust CA organizations. They are also valuable in urban offensive and defensive operations. While commanders may have sufficient combat and combat support forces, they may lack enough combat service support forces to provide the logistic support to maintain the tempo. Commanders without balanced types of forces, to include their proficiency in operating in urban environments, should consider alternatives to UO or delaying UO until proper force types are trained and available in sufficient numbers.

CASUALTIES

4-5. Casualties in UO are more likely than in operations in other environments. In urban offense and defense, friendly and threat forces often engage at close range with little space to maneuver. The urban terrain provides numerous advantages to the urban defender; higher casualties occur among troops on the offensive, where frontal assaults may be the only tactical option. Conversely, defenders with limited ability to withdraw can also suffer high casualties when isolated and attacked. Casualties can be more difficult to prevent in urban stability operations and support operations because of the dense complex terrain, the close proximity of the urban population, and the possible difficulty in distinguishing friend from foe. The potential for high casualties and the subsequent need for casualty evacuation under difficult circumstances make the positioning and availability of adequate medical resources another important consideration.

4-6. Though casualties occur in all operations, commanders recognize the likelihood of more casualties during large-scale or high-intensity UO. During the battle for Hue in 1968, for example, many company-size units suffered more than 60 percent casualties in only a few days of offensive operations. Commanders conducting urban stability operations and support operations know the casualty risk and how it relates to national and strategic objectives. While a lower risk normally exists in stability operations and support operations than in offensive and defensive operations, just one casualty may adversely impact the success of the stability or support mission. A realistic understanding of the risk and the nature of casualties resulting from UO critically affect the decisionmaking process. If commanders assess the casualty risk as high, they ensure that their higher headquarters understands their assessment and that the objectives sought within the urban area are commensurate with the anticipated risk.

MUNITIONS AND EQUIPMENT

4-7. Offensive and defensive operations in an urban environment put a premium on certain types of munitions and equipment. Forces may want to use vast amounts of precision munitions in the urban environment. At the tactical level, they will likely use more munitions than during operations in other environments. These munitions include—

or delay the overall operation. Threat forces may conduct UO with the primary purpose of causing these effects. Commanders should avoid or minimize UO that might delay or disrupt a larger operation to an unacceptable degree.

VULNERABILITIES

4-10. Commanders weigh the potential for increased vulnerabilities when executing UO. The density of the environment makes protection (safety, field discipline, force protection, and especially fratricide avoidance) much more difficult. Forces operating in a large urban area increase their risk of isolation and defeat in detail. Joint capabilities, such as air power, work less effectively to support a close urban battle than in some other environments. Thus, responding to unexpected situations or augmenting disadvantageous force ratios when applying joint capabilities is significantly more difficult. Although organized, trained, and equipped for success in any environment, the Army vulnerability to weapons of mass destruction (WMD) increases when forces concentrate to conduct UO. Commanders may consider not committing forces or limiting the size of a force committed to an urban area because of increased vulnerability to (and likelihood of) attack by WMD.

4-11. Fratricide avoidance is a matter of concern for commanders in all operations. The complex urban terrain and density of participating forces coupled with typical battlefield effects—smoke, dust, burning fires—and weather effects—fog, snow, rain, and clouds—immensely increase the potential for urban fratricide. Therefore, commanders increase emphasis on fratricide prevention measures during UO. Causes can be procedural, technical, or a combination of the two and include—

destabilizing. This consideration may limit or altogether preclude UO using Army forces.

CONSIDER ALTERNATIVES AND RISK REDUCTION MEASURES

4-14. Since UO are often high risk, commanders should consider courses of action that provide alternatives. When the objective of an urban operation is a facility, commanders should consider replicating that facility outside of the urban area. For example, a critical requirement for an airfield to sustain operations may lead commanders to consider UO to seize or secure one located in an urban area. However, if adequate resources exist, Army forces may build an airfield outside of the urban area and eliminate the need to conduct the urban operation. Similarly, logistics over-the-shore operations may be an alternative to seizing a port facility. In some situations, the objective of UO may be to protect a political organization such as a government. Relocating the government, its institutions, and its personnel to a safer area may be possible. Commanders can also design an operation to avoid an urban area. For example, if an urban area dominates a particular avenue of approach, use a different avenue of approach. This differs from isolating and bypassing because the entire operation specifically makes the urban area irrelevant.

4-15. If commanders execute UO, they assess potential hazards, and then they develop controls to either eliminate or reduce the risks to Army forces. The first means to offset risk is always to ensure a thorough understanding of the urban environment and its effects on operations by all members of the force. Other measures to bring risk to acceptable levels may include—

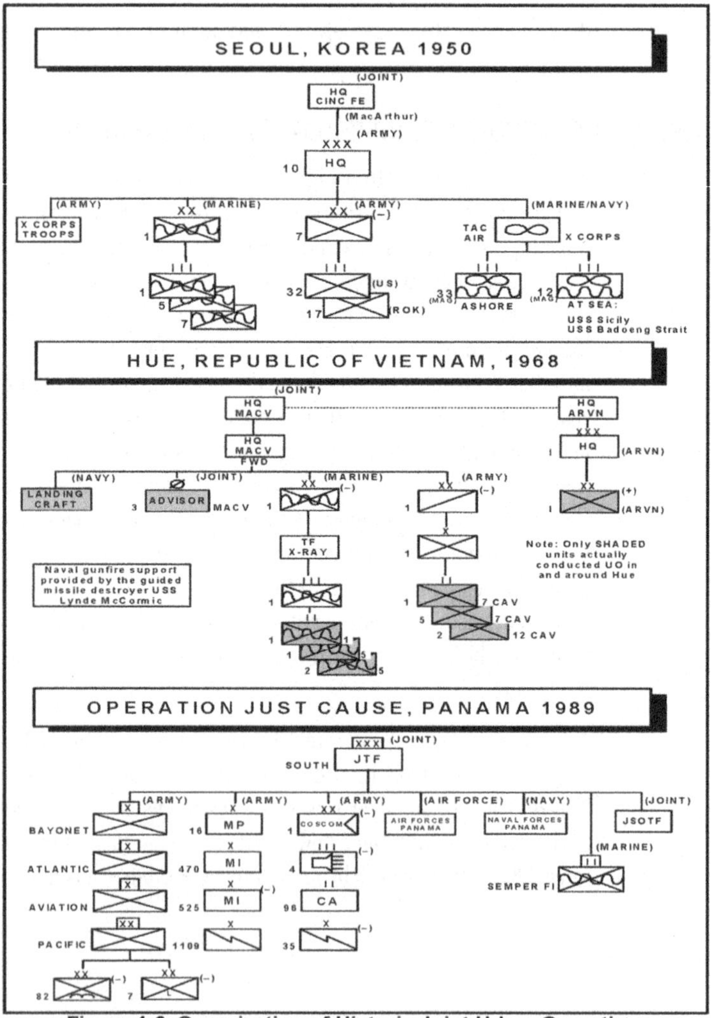

Figure 4-2. Organization of Historic Joint Urban Operations

Offense

4-19. Against a large conventional enemy in a major urban area with a large civil population present, offensive operations require the greatest commitment of Army resources. They also entail the greatest risks to Army forces and noncombatants. Within defensive or stability operations, forces may conduct tactical offensive UO, such as counterattacks to maintain the initiative or raids to eliminate elements disrupting the stability operation.

Defense

4-20. Defensive UO are generally conducted as a shaping operation within a larger major operation. These temporary operations often set conditions for successful offensive operations, stability operations, or support operations. Commanders conduct defensive UO within other types of operations to protect essential facilities in the urban area, protect flanks against counterattack, prevent the breakout of isolated enemies, or protect valuable supply bases or vulnerable convoy routes. Army forces conducting defensive UO use the environment to enhance their combat power.

Stability

4-21. Stability operations in an urban environment require offensive, defensive, and support operations, combined with other tasks unique to each stability operation. Army forces conduct urban stability operations for various reasons, including noncombatant evacuation operations, peace operations, or support to insurgencies (see Chapter 8). Urban stability operations will require an offensive capability to destroy any military capability that overtly threatens its objectives before that military threat can adversely affect the operation. Army forces employ defensive capabilities to safeguard themselves as well as secure critical places, populations, or infrastructure in the urban area. Commanders may also employ defensive capabilities to separate and protect one faction from another. Various stability tasks require urban support operations, such as distributing food or aid and protecting or assisting agencies conducting economic or humanitarian activities.

Support

4-22. Army support operations in an urban environment aid other agencies either in domestic emergencies or for humanitarian relief. Support operations require the equipment, personnel, or organizational abilities of Army forces rather than the Army's combat capabilities. In a support mission, these capabilities often involve Army transportation, medical, quartermaster, or engineer forces. Although urban support operations may seldom require combat, commanders determine if hostile threats exist that could hamper Army support operations. Defensive and offensive capabilities may be required to mitigate threats to support operations. In addition, the emergency that prompts the need for Army support operations may require stability, offensive, or defensive operations to shape the situation so units can execute support tasks.

INTEGRATION INTO LAND OPERATIONS

4-23. The commander of the major operation, after determining that urban operations are required, then integrates the urban operation into his overall operation. He does this by articulating his intent and concept for the urban operation to his subordinates. The commander of the major operation also sets the conditions for successful tactical urban operations by his subordinates. He defines ROE, focuses ISR efforts, task organizes his capabilities, ensures information superiority, designs the operational framework, and coordinates with other agencies (see FM 6-0).

CONCEPT OF THE OPERATION

4-24. The commander's concept of the operation should address all operationally important urban areas in his AO. It articulates his vision of the urban operation through directions to his staff and subordinates. Subordinate commanders address urban areas that the higher commander does not specifically address. The commander's concept discusses each urban area in terms of task and purpose (see FM 101-5). The commander also describes his vision of the situation's end state in terms of—

the key elements of the threat and environment: terrain, society, and infrastructure. To be effective, ROE are consistent throughout the force (an increased challenge in multinational urban operations), and soldiers are thoroughly trained and familiar with them.

Enemy Effects

4-28. The nature of an urban enemy affects ROE as well. Commanders consider the type of enemy weapon systems, the degree of defensive preparation, the ability to target enemy vulnerabilities with precision systems, and the ability to distinguish combatant from noncombatant.

Terrain Effects

4-29. ROE may vary according to the terrain or physical attributes of an urban area. Physical factors may drive the ROE to preclude certain types of munitions. For example, if the construction of a portion of the area is sensitive to fire, then ROE may preclude using incendiary munitions in that area. The ROE may lift this prohibition when units move into areas of mason construction. Toxic industrial chemicals or radiological contaminants in an industrial area may also affect ROE.

Societal Effects

4-30. The societal or human dimension of the urban environment will often affect ROE the most. Commanders base the ROE development on a thorough understanding of the civilian population and threat. They evaluate the loyalty of the population, its dynamic involvement in activities that affects the operation, and its size and physical location. A population that is present and supports Army forces will likely elicit more restrictive ROE than a hostile population actively supporting forces opposing the Army forces. A neutral population, not actively involved in behavior affecting Army forces, supports consideration of more restrictive ROE. In all cases, ROE conforms to the law of war. However, ROE may be much *more restrictive* than the law of war requires.

4-31. The location of the population also affects ROE. The evacuation or consolidation of noncombatants into controlled, safe areas may result in less restrictive ROE. An allied population that remains in the urban area conducting routine business in and amongst Army forces during noncombat UO will normally require the most stringent ROE.

Infrastructure Effects

4-32. Commanders consider the urban infrastructure when developing ROE. An urban infrastructure vital to current or future Army operations may dictate that commanders adjust ROE to ensure that critical elements of the infrastructure remain intact during the conduct of operations. If Army forces conduct an urban operation to capture port facilities, the ROE address damage to the key facilities that are the objective of the operation.

RESOURCE ALLOCATION

4-33. Commanders of a major operation ensure that subordinate tactical commanders have the resources necessary to conduct UO effectively. They assign appropriate forces to subordinate commanders tasked to conduct UO; support them with Army forces at the operational level; and request and coordinate their support by joint resources.

Task Organization

4-34. Task organizing subordinate units for urban operations depends largely on the nature of the operation. Some units, however, are always part of the task organization to ensure the success of UO. Infantry, CA, aviation, military police, PSYOP, military intelligence, and engineers are units required for all urban operations across the full range of Army operations. Other type forces—such as armor, artillery, and chemical—have essential roles in specific types of urban operations but are less applicable across the range of Army operations. Commanders and staffs of a major operation understand their mission, the particular urban environment in which they operate, and the general effects of the environment across the battlefield operating systems (BOS) to allocate the appropriate forces to their tactical commanders. See Chapter 5 for details.

Operational-Level Support

4-35. Commanders of a major operation also support the tactical commander with forces remaining under their direct control. These forces can include Army SOF, such as CA, PSYOP, and Special Forces, ground and air cavalry, aviation, logistics, engineers, and communications support. These forces may not be under operational control of the supported command, but their efforts are synchronized and coordinated.

Coordinating and Requesting Joint Support

4-36. Commanders of a major operation provide forces to the JFC as well as receive assets. They also coordinate for and integrate joint assets to support the tactical battle. These assets will usually include air support, such as close air support, tactical airlift, and aerial reconnaissance and surveillance. Intelligence support comes in the form of reachback to strategic and national intelligence capabilities and to space-based systems. This reachback to space assets provides reliable, robust long-range communications, environmental monitoring, and warning of enemy missile launch. Joint special operations capabilities can assist the tactical mission with special operations aviation, special reconnaissance, and direct action against high-payoff targets. Joint resources also provide the Army forces augmentation by Marine ground forces. In coastal areas, Naval forces and Coast Guard elements assist Army forces with security, sealift, and fire support. Commanders of a major operation coordinate with the JFC regarding available joint resources and their allocation. They then ensure that their efforts coordinate with and complement those of tactical Army forces in the urban area. Appendix D discusses the potential contribution of joint capabilities to Army UO.

URBAN ISR

4-37. Commanders at all levels require accurate and timely information to conduct assessments for successful urban operations. This is critical to planning and execution. Senior commanders have a large role in coordinating the urban ISR effort. National strategic sources (as well as open sources) provide most of the information that commanders and staffs require on the characteristics of the human dimension, the physical properties of the terrain, and the infrastructure. The general characteristics of these aspects of the urban environment do not change drastically over time, with one exception. Military operations or natural disasters can change physical characteristics drastically. Analysts can obtain crucial information through diligent research of intelligence databases and open sources. However, the disposition and composition of the urban threat is time sensitive and not likely to be discovered through this type of investigation. Due to the effects of the urban environment, deceptive efforts may influence the threat more easily. The urban population is dynamic and updated or confirmed as a prelude to urban operations. Surveillance and reconnaissance provide accurate and timely information regarding threat dispositions, composition and the state of the population, and the specifics of the urban terrain. Successful urban operations depend on the successful conduct of urban reconnaissance (see also the discussion of effects on the intelligence and command and control BOS in Chapter 5).

Challenges

4-38. The most significant challenge to urban ISR is physical. The physical organization and complexity of the urban terrain, both man-made and natural, challenges national strategic, operational, and tactical ISR capabilities. Commanders understand the challenges when planning and allocating time and resources to their ISR efforts. They acknowledge that subordinate commanders will face similar challenges. Therefore, commanders consider subordinate capabilities, limitations, and needs when planning, requesting, allocating, and prioritizing ISR assets and capabilities.

4-39. **Imagery Capabilities.** A significant national and strategic ISR capability is imagery. However, the structures of the urban area significantly degrade the information that imagery acquires and may make it susceptible to physical deception measures. Current imagery capabilities cannot penetrate intrasurface or subsurface areas. Yet, imagery is an excellent source regarding the arrangement and nature of many other physical aspects. It can provide significant detail of major portions of the infrastructure. Imagery can also reveal what may be happening in structures through detailed study of patterns and other exterior indicators. Yet, the bulk of a skillful threat's forces, well positioned and concealed inside or underneath structures in the urban area, are largely immune from rapid detection by overhead imaging systems. The volume of movement in an urban area will itself provide a degree of camouflage and increase the difficulty of employing pattern analysis. The success in 1999 of the Yugoslavian army concealing heavy forces when confronting NATO indicates the limits of these assets to penetrate an urban area.

4-40. **Electronic Capabilities.** The physical attributes of the urban area also diminish the effectiveness of electronic ISR capabilities. Buildings and other structures significantly disrupt radio communications in an urban area. Buildings not only make tactical radio communications difficult for the user, they also make them difficult to locate, intercept, and jam. The range and clarity of frequency modulation (FM) signals significantly diminish when antennas are located inside buildings or when buildings block line of sight between the source and receiving station. To mitigate this effect, detection capabilities often move closer to the transmission source. Without losing tactical surprise and increasing risk, units cannot effectively use many electronic detection and surveillance capabilities until urban combat is imminent or perhaps already begun. Thus, the threat's vulnerability to compromise by means of his FM and other wireless communications in an urban environment is much less than in many other environments.

4-41. **Human Capabilities.** The limits on imagery and electronic ISR capabilities place a premium on human-based visual reconnaissance. Commanders have three types of human reconnaissance assets to augment electronic reconnaissance resources: special reconnaissance, conventional combat reconnaissance, and human intelligence (HUMINT) gathered by military intelligence from individuals. The urban environment poses several challenges to these capabilities.

4-42. The urban area challenges special reconnaissance in several ways. First is the access the urban area. Although avenues of approach may be numerous, *concealed* avenues of approach into a defended urban area may be limited and thoroughly covered. Air access is also more difficult because aircraft are detected more easily, airspace is smaller, drop and landing zones are limited or not secure, and more air defense systems probably exist. Still, special reconnaissance efforts to penetrate the urban area can be successful using unconventional techniques including high-altitude low-opening parachutes or underwater penetration.

4-43. Special reconnaissance then faces a second challenge: moving in and identifying targets in the urban area. Stealth movement in an occupied urban area is exceptionally difficult. Repositioning to new or alternate positions is also dangerous. The soldiers' ability to conceal themselves among the civil population can mitigate some of these challenges but includes inherent risks of a different nature. Also difficult is establishing observation positions that provide a field of view of several targets.

4-44. Finally, special reconnaissance may face navigational and reporting challenges. Special reconnaissance's ability to locate themselves and communicate critical locations and routes are challenged by—

4-45. Conventional reconnaissance faces many of the same challenges as special reconnaissance. Conventional reconnaissance also may lack the advantage of surprise and the special equipment and training that provides special reconnaissance stealth capability. Conventional reconnaissance is not likely to operate undetected by the civilian population. Given the constraints discussed above on other sources, conventional reconnaissance units will likely begin their mission with much less information than they would have on threat dispositions in a less complex environment. Commanders may choose to have their reconnaissance elements fight for information in the urban area. While this high-risk option is more favorable under fluid conditions, it can be used at any time. It requires careful planning, rehearsal, and formulation of information requirements.

4-46. Human intelligence may be one of the most valuable sources for information regarding the situation inside an urban area. HUMINT may take advantage of the proximity and large numbers of potential informants to gather information about threat activities and capabilities. It is especially valuable because it can address all elements of the environment. HUMINT sources can describe political and religious nuances significant to commanders. Such information is useful for insights regarding the human dimension but extremely difficult to obtain from other means. This intelligence also can describe the infrastructure relating essential details of how the infrastructure functions. Obtaining good HUMINT requires skilled interrogators and linguists. Commanders know and account for some of the possible shortcomings of HUMINT:

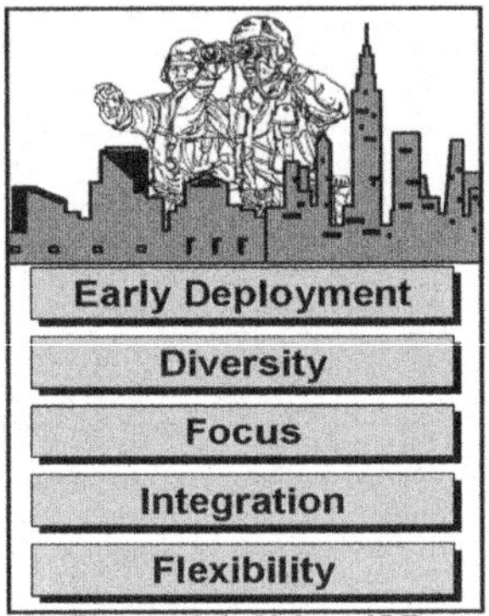

Figure 4-3. Urban ISR Considerations

4-49. Limited national, strategic, and operational imagery intelligence (IMINT) and signals intelligence (SIGINT) capabilities are requested. If they are approved, they are tasked and deployed or repositioned to begin urban ISR operations. This takes time. Spacing the ISR effort over time permits the analysis of the information or data as it is received. Such time also permits subsequently refining the ISR effort before all assets are committed.

4-50. SOF or conventional units will require significantly more time to execute reconnaissance missions and maintain an acceptable survivability rate. Urban reconnaissance operations require additional time for stealthy insertion into the urban area. IMINT and SIGINT capabilities are used to identify possible locations of high-value targets and corresponding observation positions; this helps minimize time-consuming and high-risk repositioning in the urban area. Again, reconnaissance units may require extensive time to observe from observation positions for indicators of threat activity and disposition and identify patterns.

4-51. As conventional combat forces prepare to commit to the urban area, conventional reconnaissance precedes their actions. Conventional reconnaissance will often be a slow and methodical effort. Such forces need time to reconnoiter the interior of structures for snipers and other small threat teams. They also need time to deploy and destroy snipers and small delaying elements and to breach harassing obstacles. If necessary, they need time to mass the combat power necessary to fight through security forces and continue reconnaissance.

4-52. **Diversity.** No single ISR capability can solve the riddle of the urban defense. The only way to successfully gain a thorough common operational picture of the complex urban terrain—so commanders can focus combat power on decisive points—is to employ diverse ISR capabilities. These capabilities will each contribute pieces of relevant information to permit identifying operational objectives and leveraging tactical combat power to achieve those objectives quickly. Higher-level commanders know that tactical reconnaissance capabilities alone often cannot provide all the tactical information required for success at lower echelons.

4-53. Using diverse capabilities challenges the threat's ability to defeat the friendly ISR effort. A threat who focuses on minimizing his vulnerability to satellite imagery may increase his reliance on communications and thus his vulnerability to SIGINT. At the same time, he may decrease his ability to detect the actions of ground reconnaissance units. A threat that actively campaigns to detect ground reconnaissance may make himself more vulnerable to SIGINT and IMINT.

4-54. Diverse capabilities also facilitate the tactical ISR effort. Tactical reconnaissance units often consist of small dismounted teams and small combined arms teams with a dismounted element and an armor-protected mounted element. Engineers and breaching capability are essential to the combined arms reconnaissance effort. The teams' movements are synchronized and coordinated with other assets, such as unmanned aerial vehicles (UAVs) and air cavalry reconnaissance. These teams use several movement techniques including infiltration, with the primary objective of conducting zone reconnaissance along key axes that support brigade and battalion actions against decisive points. To accomplish this mission, reconnaissance reconnoiters the proposed routes and alternate approaches. This supports deception and contingency planning. Infiltration of dismounted reconnaissance is made easier when a threat focuses on combined arms reconnaissance teams. Aerial reconnaissance, such as air cavalry and UAVs, provides early warning of threat elements to ground reconnaissance, identifies obstacles and ambush sites, and helps select the routes for ground reconnaissance. Air elements may also reduce the mobility of counterreconnaissance forces.

4-55. **Focus.** Another key to successful ISR is the ability to focus the assets on commander's critical information requirements (CCIR). This focus begins with the mission and the commander's initial planning guidance. It is incrementally refined throughout planning and execution as each ISR effort provides information and permits more specific focus in subsequent efforts. The size and complexity of the urban environment require that the ISR effort center strictly on decisive points or centers of gravity (COGs). Therefore, the overall ISR effort will have two major focuses. The first is to confirm and develop information on the decisive points and COG. The second is the approaches to the decisive points and COG. The first focus will likely drive ISR in support of major operations. The second focus will likely provide the impetus for tactical ISR efforts. For example, special operations reconnaissance might focus on a major command center that controls the entire urban area and that is one of a corps CCIR. Tactical reconnaissance might focus on the nature of the defense along a particular avenue of approach to the objective.

4-56. **Integration.** Another important aspect of urban ISR is integration. All reconnaissance capabilities provide both distinctive information as well as information that confirms and adds to that coming from other sources. Essential to urban ISR is the link between all of these sources, either directly or through an integrating headquarters.

4-57. ISR operations are vertically and horizontally linked. Vertical links ensure that ISR operations among the various levels of command are complementary and that the information flow between these levels is rapid. Horizontal links ensure that forces operating in close proximity (particularly adjacent units), where areas of interest overlap, can rapidly share results of their individual ISR efforts. Together, this helps ensure that all Army forces share a common operational picture and permits the greatest flexibility and survivability of ISR resources.

4-58. ISR operations also are integrated into the planning system, especially the targeting process. As part of targeting, positioned reconnaissance and surveillance elements may become the trigger and terminal control for applying precision fires when appropriate and after considering the risks of compromise of the position or platform.

4-59. **Flexibility.** The urban ISR effort is more flexible than in other operations. This flexibility permits the ISR effort to meet unforeseen circumstances and to deal with the challenges of the urban environment. As indicated previously, the urban environment is particularly difficult to penetrate. The practical effects of this characteristic are that—

4-62. IO are executed using core and supporting elements and related activities (see Figure 4-4 and FM 100-6). The elements of IO are employed in either an offensive or defensive role. Many elements of IO are not affected differently in an urban environment from any other environment. The following sections outline some IO considerations unique to urban operations.

Figure 4-4. IO Elements and Related Activities

4-66. The positive influence created by PSYOP is often essential to developing an effective HUMINT capability particularly in an urban area where many civilians actively or passively support the threat. Persuading and influencing a few to support friendly forces may pay great dividends. These few supporters may allow Army forces to penetrate the urban area and obtain essential information. Such information can apply to threat capabilities, threat intentions, and even the urban environment itself.

4-67. PSYOP, combined with other elements of offensive IO, aid in isolation of a threat—a critical shaping action for any urban operation. For example, commanders may use PSYOP to inform civilians about new food distribution points located away from urban combat operations. This action supports the UO fundamental of separating combatants from noncombatants and helps to further isolate the threat (both physically and psychologically) from the civilian populace. Aside from projecting a positive image of friendly forces over threat forces, PSYOP also isolates the threat. These operations identify and exploit ethnic, cultural, religious, and economic differences between the elements of the civilian populace and threat forces as well as the differences among supportive and unsupportive civilian factions. The complexity of the urban environment enables quick changes in opinion or attitude. Commanders continually evaluate the results of PSYOP for mission relevance.

Counterpropaganda

4-68. Because propaganda is aimed at both combatants and noncombatants, UO are especially concerned with its use. Propaganda can rapidly and dramatically affect the attitudes of the urban population and will probably occur after urban operations have begun. Thus, it can create situations in the human dimension of the environment quite different from those discovered in the pre-operations assessment. Counterpropaganda is, therefore, essential to urban operations. To negate, deflect, or destroy the threat's propaganda capability, counterpropaganda requires—

differential necessary to achieve success. Commanders tailor urban deception plans to the specific urban area, paying close attention to the societal characteristics of the target population.

Electronic Warfare

4-70. Electronic warfare (EW) includes all actions that use electromagnetic or directed energy weapons to control the electromagnetic spectrum or to attack a threat. Conducting EW in urban areas seeks to achieve much the same results as in other environments. A major consideration in urban areas is collateral effects on portions of the urban infrastructure that rely on the electromagnetic spectrum for service. Thus, precision is a major factor in planning for EW operations. For example, EW attacking a threat's television broadcasts avoids affecting the television broadcasts of neutral or friendly television. Likewise, EW attacking military communications in a large urban area avoids adversely affecting the area's police and other emergency service communications. Urban offensive and defensive operations will have the least restrictions on EW operations while urban stability operations and support operations may have significant constraints on using EW capabilities.

Computer Network Operations

4-71. Computer network operations (CNO) include computer network attack (CNA), computer network defense (CND), and computer network exploitation (CNE). CNO are not applicable to units at corps and below. Echelons above corps (EAC) units will conduct CNA and CNE. If tactical units require either of these network support, they will request it of EAC units.

4-72. **Computer Network Defense.** In urban operations, CND will require extreme measures to protect and defend the computers and networks from disruption, denial, degradation, or destruction. The nature of the urban environment and configuration of computer networks provides the threat with many opportunities to interdict local area networks (LANs) unless monitored by military forces. LANs controlled by military forces are normally more secure than the civilian infrastructure. Commanders prepare for opportunities by the threat to insert misinformation.

4-73. **Computer Network Attack.** Considerations regarding the execution of CNA in urban operations are similar to those of EW: CNAs that do not discriminate can disrupt vital civilian systems. However, possible adverse effects on the civilian infrastructure can be much larger—potentially on a global scale. In the short term, CNAs may serve to enhance immediate combat operations but have a debilitating effect on the efficiency of follow-on urban stability operations. Because of these far-reaching effects, tactical units do not execute CNA. CNA is requested of EAC units. EAC units will receive all requests from lower echelons, carefully consider second- and third-order effects of CNA, and work to ensure its precise application.

4-74. **Computer Network Exploitation.** CNE is an enabling operation and intelligence collection to gather data from target or adversary automated INFOSYS or networks. Tactical units do not have the capability for CNE. CNE contributes to intelligence collection at EAC. In UO, CNE will be centrally controlled.

Information Assurance

4-75. Information assurance in UO takes on an added dimension. As with other operations, availability of information means timely, reliable access to data and services by authorized users. In UO, the timeliness of information may be restricted because structures block the transmission waves. The need for retransmission facilities will overwhelm the signal community. The reliability can be questioned because of the blockage between units and communications nodes. Unauthorized users may intercept the communications and input misinformation or disinformation. Commanders protect the integrity of all information from unauthorized changes, including destruction. INFOSYS with integrity operate correctly, consistently, and accurately. The authentication of information may be accomplished by sophisticated electronic means. However, it is more likely that communications-electronics operating instructions authentication tables will authenticate the information. Commanders consider the confidential nature of all information in UO. The G6 protects the information from unauthorized disclosure. Information being passed cannot be repudiated. The density of the infrastructure in urban areas may inhibit receipt by the intended individual or unit. The sender may have no means to determine if the message was received.

Counterdeception

4-76. In UO, threat forces can easily accomplish deception operations. The force that controls the area above and below ground will have freedom of movement. Deception aimed at friendly commanders will cause them to deploy combat power at the wrong place and the wrong time. Counterdeception by friendly commanders will identify and exploit threat attempts to mislead friendly forces. Counterdeception is difficult. Cultures of certain rhetoric and actions are more predisposed to deception than others. Knowing a threat's previous deception methods is important. Dismissing tactical indicators because they conflict with preconceptions may allow a hostile deception operation that plays on the preconception to succeed.

Physical Destruction

4-77. Physical destruction includes those actions—including direct and indirect fires from air, land, sea, space, and Special Forces—taken with, to augment, or supplement IO actions. Like many other IO elements, major concerns with employing physical destruction in UO are precision and follow-on effects. Thus, commanders using physical destruction to support IO adhere to the same constraints as all other fires.

Counterintelligence

4-78. Counterintelligence (CI) in the context of IO focuses on detecting threats against INFOSYS. The urban environment, particularly in stability operations and support operations, is ideal for espionage, other intelligence activities, sabotage, or assassination. Threats can approach, conduct reconnaissance, and escape under the concealment of the urban terrain and population.

Civil-Military Operations

4-79. Civil-military operations (CMO) are a critical aspect of virtually every urban operation and are included here as a closely related activity of IO. CMO activities enhance the relationship between military forces, civilian authorities, and the urban population. They promote the development of favorable emotions, attitudes, or behavior. CMO range from support to combat operations to assisting in establishing political, economic, and social stability. Chapter 9 has a more detailed discussion of CMO and CA units. However, because of its criticality to UO, CMO and its effects are thoroughly integrated throughout this manual.

Public Affairs

> *Four hostile newspapers are more to be feared than a thousand bayonets.*
>
> Napoleon Bonaparte

4-80. Another related activity to IO is public affairs (PA). PA influences urban operations by transmitting information through the media to internal (in urban Army forces as well as in the urban civilian populace) and external audiences. At higher levels of command, PA can help maintain popular national support for the urban operation by clarifying the links between strategic goals and operational objectives. At both the operational and tactical levels, it links Army units, the urban inhabitants, the US and international public, and the media. PA can help determine potential media issues that may influence planned UO. It can also aid commanders in assessing the impact of UO on the environment (particularly its citizens) and other agencies and organizations operating in the urban area. PA also helps to counter rumors, uncertainty, fear, loneliness, confusion, and other factors that cause stress (to both soldiers and civilians) and undermine effective UO. If the populace does not understand the mission, false expectations may be created that Army forces may not be able to meet. PA can help prepare the American public for the possibility of high casualty rates. Overall, PA supports urban commanders in their goals to achieve information superiority and preserve public support.

4-81. PA does not distort, direct, or manipulate information. Its effectiveness stems directly from establishing and maintaining credibility with the urban population and media. Commanders synchronize PA with the integral elements of IO (particularly PSYOP and counterpropaganda) to ensure that all Army sources send only one message. Urban commanders plan for the media and integrate PA into their decisionmaking and (through IO) targeting processes.

4-82. The density of information sources and reporters in UO ensures that all Army activities will be subject to media and public scrutiny. Many reporters will congregate in cities for their own comfort and take advantage of established communications networks. Urban areas are densely populated and, together with Army forces and NGOs operating there, will present the greatest number of human-interest stories. The local urban or host-nation media, however, will often have their own agendas developed over a longer period of time. This local media will also have a greater influence over the urban population than the international media. The indigenous media may not follow international norms. Commanders are responsible to understand the media (particularly the local media), its role, and its potential influence. They cannot allow themselves to be intimidated by it. Commanders support open and independent reporting and grant access to their units as early and as far forward as the situation permits.

with the SOF units operating in their AOs. It primarily deconflicts conventional and SOF targets, positions, and missions. The synchronization and unity of action necessary between conventional and SOF in an urban AO may still require the Army force headquarters to further coordinate SOF integration through the JTF commander. The special operations coordination element (SOCOORD) is the ARSOF element within the Army corps or Marine expeditionary force (MEF) G3 section responsible for coordinating special operations requirements. As an integral part of the corps or MEF staff, the SOCOORD provides a focal point for SOF command, control, communications, computers, and intelligence structure to synchronize special operations activities in support of corps missions.

COORDINATION WITH OTHER AGENCIES

4-87. The population density of the urban environment, its economic and political importance, and its life-supporting infrastructure attracts many types of organizations. These organizations include—

urban areas. CMOCs may be organized in various ways and include representatives from as many agencies as required to facilitate the flow of information among all concerned parties. Commanders still ensure that force protection and OPSEC requirements are not compromised. Effective CMOCs can serve as clearinghouses for the receipt and validation of all civilian requests for support, can aid in prioritizing efforts and eliminating redundancy, and, most importantly, can reduce wasting the urban commander's scarce resources.

Liaison Officers

4-89. Liaison officers (LNOs)—sufficiently experienced and adequately trained in liaison duties and functions—are necessary to deal with the other agencies that have interests in the urban area. Army LNOs work with the lead agency or other organizations that the commander has identified as critical to mission success. Together they work to rapidly establish unity of effort and maintain coordination, often before a CMOC is established. The additional coordination afforded by the physical presence of LNOs within these organizations may be required even after the CMOC is fully functional. When commanders lack enough LNOs to meet requirements, they prioritize and often assign a single LNO to several organizations. That LNO will then share his time and presence to those organizations based on the situation and his commander's guidance.

Commander's Personal Involvement

4-90. Overall, establishing a close relationship with other agencies will often be a major, positive factor in successful mission accomplishment, particularly in urban stability operations. Commanders that develop a direct and personal relationship with the leaders and staff of other agencies can often avoid conflict, win support, and help eliminate the "us versus them" mentality that frequently frustrates cooperation among Army forces and civilian organizations.

Chapter 5

Foundations for Urban Operations

Utilities such as electricity and water are as much weapons of war as rifles, artillery pieces or fighter aircraft. . . . In the case of Manila, where there was a noncombatant, civilian population of one million in place, it was the attacker's aim to capture the utilities which the defender planned to destroy.

The Battle for Manila

Commanders conducting major urban operations (UO) use their ability to visualize how doctrine and military capabilities are applied within the context of the urban environment. An operational framework is the basic foundation for this visualization. In turn, this visualization forms the basis of operational design and decisionmaking. To accurately visualize, describe, and direct the conduct of UO, commanders and their staffs understand the basic fundamentals applicable to most UO. They also understand how the urban environment affects the battlefield operating systems (BOS) and the tactical urban battle.

URBAN OPERATIONAL FRAMEWORK

5-1. Army leaders who have an urban area in their area of operations (AO) or are assigned missions in an urban area follow an urban operational framework. They identify the portion of the urban area essential to mission success, shape the area, precisely mass the effects of combat power to rapidly dominate the area, and then transition control of the area to another agency. This framework divides into four essentials: *assess*, *shape*, *dominate*, and *transition*. These four components provide a means for conceptualizing the application of Army combat power and capabilities in the urban environment. The Army framework modifies the joint urban operations framework

CONTENTS

(understand, shape, engage, consolidate, and transition) to further clarify the JUO concepts within the context of Army capstone doctrine found in FM 3-0. The framework for joint urban operations (JUO) provides the joint force commander a framework for planning and conducting JUO. FM 3-0 provides Army commanders with the operations process that provides a framework for planning, preparation, execution, and continuous assessment. Army capstone doctrine, supported with the Army UO framework, is fully compatible with the concepts and purpose of the JUO framework.

5-2. The urban operational framework assists commanders in visualizing urban operations. This framework is simply an aid to the commander. Commanders combine the framework with—

ASSESS

5-3. Assessment is the continuous monitoring—throughout planning preparation, and execution—of the current situation and progress of an operation, and the evaluation of it against criteria of success to make decisions and adjustments (FM 3-0). Commanders use visualization as their assessment method, staff officers use staff estimates, and all use the intelligence preparation of the battlefield (IPB) process. Commanders and staffs begin the assessment process by observing and then collecting information about the situation. They observe and learn about the urban environment, and factors of METT-TC—mission, enemy, terrain and weather, troops and support available, time available, civil considerations. They use intelligence, surveillance, and reconnaissance means; information systems (INFOSYS); and reports from other headquarters, services, organizations, and agencies. Then they orient themselves to the situation and achieve situational understanding based on a common operational picture (COP) and continuously updated CCIR. Largely, the ability to rapidly and accurately assess the situation contributes to the commanders' abilities to seize, retain, and exploit the initiative during UO.

Disproportionately Critical

5-4. The Army operations process requires continuous assessment; it precedes and guides every activity. In UO, however, assessment is disproportionately critical for several reasons. First, each urban environment is unique. Other environments can be studied and their characteristics quantified in a general manner with accuracy. This is fundamentally not true of different urban areas. The characteristics and experience in one urban area often have limited value and application to an urban area elsewhere. This characteristic sets UO apart from operations in other environments.

Extremely Dynamic

5-5. The urban environment is also extremely dynamic. Either deliberate destruction or collateral damage can quickly alter physical aspects of the urban environment. The human aspect is even more dynamic and potentially volatile. A friendly civil population, for example, can become hostile almost instantaneously. These dynamics (combined with initial difficulty of understanding and describing this unique environment) make it difficult for commanders and staffs to initially develop and maintain a COP and establish situational understanding. Furthermore, public reaction to media coverage of the urban operation and political changes influence national objectives and overall strategy. Such changes can affect the basic nature of an operation, especially after it has commenced. Anticipating these potential effects and developing appropriate branches and sequels based on an accurate assessment often determines how quickly commanders can achieve the desired end state.

Risk Assessment

5-6. As in any environment, UO pose both tactical and accident risks. However, the level of uncertainty, ambiguity, and friction can often be higher than that of many other environments. Such challenges increase the probability and severity of a potential loss due to the presence of the enemy, a hostile civilian group, or some other hazardous condition within the urban environment (see Necessity of Urban Operations in Chapter 4). Therefore, commanders—

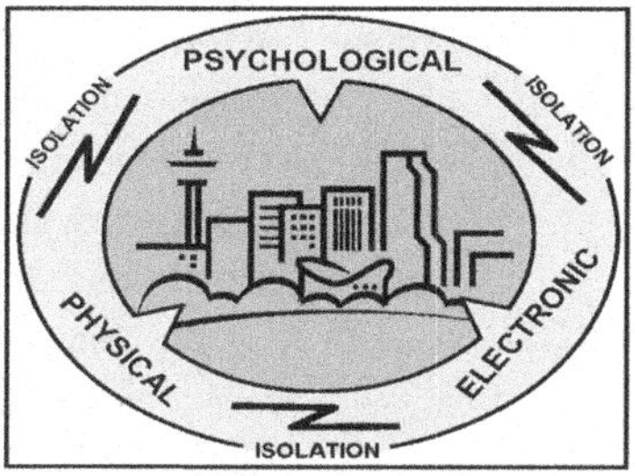

Figure 5-2. Urban Isolation

key shaping action that affects UO. It applies across the range of Army operations. Most successful UO have effectively isolated the urban area. Failure to do so often contributed to a difficult or failed UO. In fact, the relationship between successful isolation and successful UO is so great that the threat often opposes isolation actions more strongly than operations executed in the urban area. In some situations, the success of isolation efforts has been decisive. This occurs when the isolation or imminent isolation of the urban area compels a defending enemy to withdraw or to surrender before beginning or completing decisive operations. In UO that are opposed, Army forces attempt to isolate the threat three ways: physically, electronically, and psychologically (see Figure 5-2).

5-10. **Physical Isolation.** In offensive UO, physical isolation keeps the threat from receiving information, supplies, and reinforcement while preventing him from withdrawing or breaking out. Conversely, a defending Army force attempts to avoid its own physical isolation. Simultaneously, this force conducts operations to isolate the threat outside, as they enter, or at selected locations in the urban area. Physical isolation can occur at all levels. In many situations, particularly major theater war (MTW), the commander of a major operation may attempt to isolate the entire urban area and all enemy forces defending or attacking it. At the tactical level, forces isolate and attack individual decisive points. In stability operations, physical isolation may be more subtly focused on isolating less obvious decisive points, such as a hostile civilian group's individual leaders. In many operations, isolation may be temporary and synchronized to facilitate a decisive operation elsewhere. To effectively isolate an urban area, air, space, and sea forces are necessary in addition to the capabilities of ground forces.

5-11. **Electronic Isolation.** Electronic isolation is achieved through offensive information operations (IO). Electronic warfare (particularly two of its components: electronic warfare support and electronic attack) and computer network attack are critical to electronic isolation (see FM 100-6 and Information Operations in Chapter 4). At the operational level, offensive IO aims to quickly and effectively control the information flow into and out of an urban area. This isolation separates the threat's command and control (C2) system - in the urban area from its operational and strategic leadership outside the urban area. Offensive IO also focuses on preventing the threat from communicating with civilians through television, radio, telephone, and computer systems. At the tactical level, IO aim to isolate the threat's combat capability from its C2 and leadership within the urban area, thus preventing unity of effort within the urban area. Defensive IO are key to preventing isolation of friendly forces defending in an urban area.

5-12. **Psychological Isolation.** Psychological isolation is a function of public affairs, physical actions, electronic warfare, and other forms of IO, especially military deception and psychological operations. Psychological isolation denies the threat political and military allies. It separates the enemy or hostile civilian group from the friendly population, nongovernmental organizations (NGOs) operating in the urban area, and from political leaders who may consider supporting Army forces. Psychological isolation destroys the morale of individual enemy soldiers or hostile civilians. It creates a feeling of isolation and hopelessness in the mind of the threat. It undermines the confidence of the threat in their leadership. On the other hand, IO, as well as the disciplined conduct of Army personnel, can help to forge legitimacy for Army operations. In stability operations, psychologically isolating the threat results in the friendly urban population and NGOs positively supporting Army operations.

Other Shaping Actions

5-13. Other shaping actions can include the proper sequencing and deployment of forces, reconnaissance operations, and force protection. These actions contribute equally to the success of any urban operation. Commanders understand how the urban environment affects their ability to accomplish these shaping actions. However, civil-military operations (CMO), another closely related activity of IO, are important to shaping the urban battlespace for decisive operations. The specific civil-military task can vary greatly and may include affecting a cooperative relationship with the civil political system, protecting portions of the civil population or infrastructure, or establishing refugee camps or safe areas for noncombatants. This is most true in stability operations and support operations. Successful CMO also can contribute to the psychological isolation of the threat. (See Civil-Military Operations in Chapters 4 and 9 for more detailed discussions.)

Training and Education

5-14. Finally, Army commanders know that critical shaping actions often occur prior to the urban operation in the form of professional education and training. Commanders can enhance training through joint, interagency, multinational, and combined arms exercises and effective rehearsals. Capabilities and competencies of units include—

capabilities to achieve maximum effects. Army forces dominate a situation when they have fulfilled all mission requirements and established preeminent military control over the threat, geographical area, or population. Achieving domination in a specific urban operation depends, of course, on the situation and the assigned mission.

Offense: Attack Decisive Points

5-16. In urban offensive operations, forces achieve dominance by successfully striking at the enemy's center of gravity using multiple offensive actions from unexpected directions and throughout all dimensions. Army forces aim to dominate identifiable decisive points. Successful efforts against decisive points lead to effects on the center of gravity. The center of gravity will differ in each offensive situation. It may be an individual enemy leader, the enemy's combat power, the enemy's communications capability, or a physical structure of cultural, political, or economic significance.

Defense: Deny Vital Functions and Critical Infrastructure

5-17. In urban defensive operations, domination translates into denying the enemy control of the vital functions and critical infrastructure of the urban area. Forces achieve this by leveraging the defensive advantages of the urban terrain, defending essential areas in depth, using economy of force in nonessential areas, controlling the enemy direction of attack with natural and man-made obstacles, and retaining the initiative through counterattacks.

Stability and Support: Apply Innovation and Imagination

5-18. The ability to dominate in urban stability operations hinges on the type of stability operation commanders execute. In a noncombatant evacuation operation, forces limit domination to finite geographic areas and times. In contrast, a peace operation may require domination of a large urban area for an extended time. In this operation, dominate is defined as using the array of Army capabilities to create specific conditions among the belligerents. Thus, the techniques used for domination in stability operations vary according to the situation and as situations mature during long-term operations.

5-19. In urban support operations, dominating the situation may require innovative and subtle application of Army capabilities. Since Army forces usually support other agencies that lead the operation, achieving domination results from carefully and discretely applying Army capabilities to the tasks assigned by the lead agency. In a humanitarian relief situation, Army forces may be tasked to transport supplies in the urban area. Domination of this activity then becomes the goal of Army forces and may be achieved by providing, managing, and protecting transportation assets.

TRANSITION

5-20. When planning UO, commanders ensure that they plan and prepare for transitions. Transitions are movements from one phase of an operation to another. They involve significant changes in the type of operation, concept of the operation, mission, situation, task organization, forces, resource allocation and support arrangements, or C2. Transitions occur in all operations, but in UO they occur with greater frequency and intensity, are more complex, and often involve agencies other than US military organizations. For example, a successful attack may transition to a defend mission that includes not only defense tasks but also stability tasks. Unless planned and executed effectively, transitions can reduce the tempo of the urban operation, slow its momentum, and cede the initiative to the threat.

Mental and Physical Preparation

5-21. Transitions occur as conditions warrant. They can be carefully planned and controlled, or they can be quick and dramatic, such as the swift transformation of a stability operation into offense or defense. Units prepare mentally and physically to address rapid transitions. Accordingly, plans include branches and sequels that address anticipated or possible transition points. When the dominant type of operation changes from an offense to stability, the types of units originally conducting the UO may no longer be appropriate. A large mobile reserve may permit increased flexibility to react to unplanned transition requirements. Operations in one part of an urban area may transition before operations in a different part of the same urban area. This will require commanders to execute various types of operations and associated tasks simultaneously.

Transition to Legitimate Civilian Authorities or Agencies

5-22. In UO, a distinct aspect of transition is the requirement to quickly and efficiently transition the major portions of Army responsibilities to civil agencies. Some tasks to which units will transition are not traditional combat tasks but rather stability tasks more closely associated with CMO. In stability operations and support operations this is often a near-term critical mission objective. In these operations, commanders aim to alleviate the circumstances requiring Army forces and ensure that other civilian agencies assume the functions provided by Army forces. In combat operations, civilian agencies quickly resume specific support activities—such as providing sanitary services, food services, law enforcement, and health services—because of their high demand on Army resources.

Clearly Visualize and Describe the End State

5-23. Army UO conclude when Army forces depart and have no further mission requirements in the urban area. At the outset, commanders visualize and describe the intended end state of a unit's execution of UO. Commanders then clarify and update this visualization as the political or strategic situation is refined or changes. This enables subordinate units to identify likely transitions and ensures that current operational planning takes into account second- and third-order effects. As long as an active Army AO contains an urban area, some type of urban operation will exist. After urban combat successfully ends, combat forces may move on. Support forces conducting sustaining operations may then occupy the area and continue to conduct a different form of UO.

Applying the Urban Operational Framework
Panama – December 1989

The US conducted OPERATION JUST CAUSE in December 1989 to remove the illegal ruler of Panama, Manuel Noriega, and to restore that country to a democracy. It also conducted the operation to ensure the safety of a substantial number of US personnel as well as the security of US interests in Panama. The major focus of JUST CAUSE was in Panama City, the country's capital. Most operations occurred in this large urban area, one of the numerous smaller urban areas, or the urban-like military bases. These bases proliferated the AO and were directly linked to operations in the capital city. This successful operation illustrates how commanders can apply the urban operational framework to visualize, describe, and direct the critical aspects of urban operations.

Assess

The synchronization achieved during the operation may have obscured the challenges faced in the initial assessment process in Panama. However, it was not as simple as it may have seemed. Using the framework of the urban environment, US forces required details of the physical characteristics of the environment, the infrastructure, and the human dimension including the capabilities of the Panamanian military.

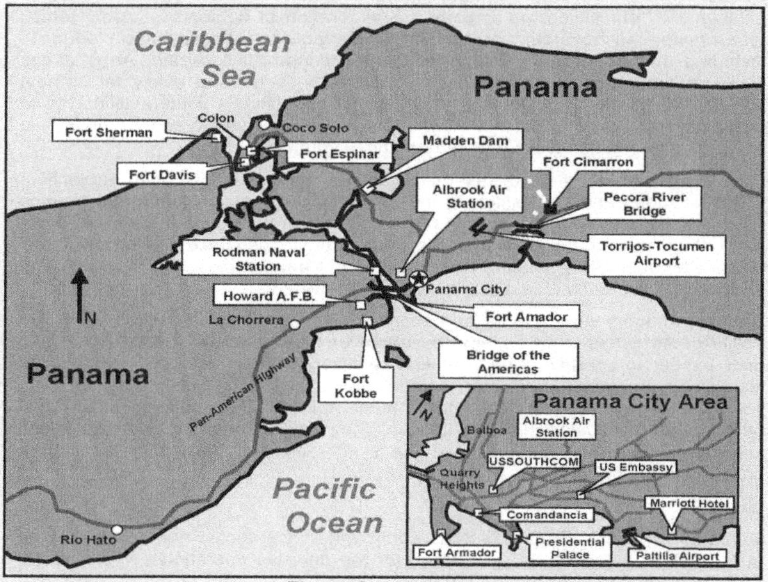

Figure 5-3. Panama

Because Army forces had a long history in Panama, commanders clearly understood the physical challenges and layout of critical urban areas (see Figure 5-3 on page 5-9), particularly Panama City. They also understood how the infrastructure in each urban area functioned and which parts would be key to success. Examples of key portions of the infrastructure included the Madden Dam, which controlled the water flow through the Panama Canal, and the Cerro Azul television tower, which was the main Panamanian broadcast tower.

Collecting information and developing intelligence on the human elements of the urban environment was critical to operational success and a challenge. Because the Panamanian Defense Force (PDF) had traditionally been an ally of the US, Army forces did not have a systemic database that adequately depicted their order of battle and their true capabilities. Additionally, much of the situation in Panama was colored in political terms making it more difficult for traditional military sources to evaluate the status of PDF forces. For example, Army planners needed to know if PDF military units (when faced with a formidable US force) would fight at all for Noriega and if they did fight, how hard and long would they fight. The answers depended largely on their political loyalty to Noriega and on the individual loyalty of the unit officers to the Panamanian president. Thus, Army commanders needed to understand the military characteristics of PDF units and their political affiliations and tendencies.

Because transition from combat to noncombat tasks would be critical to achieving all objectives, particularly the restoration of democracy, Army forces also needed an accurate assessment of the political opposition to Noriega—including that opposition's capabilities and vulnerabilities. Again, Army forces were required to make assessments outside those needed solely for combat operations. Ultimately, assessing the political opposition's vulnerabilities led to assigning Army units to protect them throughout the operation so that they could serve as a foundation for a new democratic government.

Finally, the commander's assessment included an evaluation (often subjective) of the attitudes and disposition of the Panamanian people. Human intelligence (HUMINT) was the primary source of information on the population. Army forces had good access to the population because of their close proximity and historical ties to Panama. Many soldiers were married to Panamanians, and the Army had total access to local media and to prominent individuals.

National imagery and special operations forces (SOF) also contributed to the ability of Army forces to assess the urban environments of JUST CAUSE. All units executing operations had detailed satellite photos of objective areas. Additionally, key objectives were placed under SOF surveillance well in advance. This surveillance revealed unit readiness, vulnerabilities, detailed disposition, and other patterns critical to mission success. The combination of the two capabilities allowed units to plan and achieve the synchronization necessary for such a complex urban operation.

Shape

During OPERATION JUST CAUSE, commanders conducted numerous shaping operations to establish the conditions for the decisive operations. Many operations were designed to control information, such as an assault on the Azul television tower identified during the assessment of the infrastructure. Planners designed many shaping operations to isolate various garrisons and PDF units.

An example of tactical isolation was the plan for the Pacora River Bridge to prevent reinforcements from reaching the garrison at Torrijos-Tocumen Airport.

Operational isolation was achieved through the Ranger Regiment's and 82nd Airborne Division's assault on targets at Rio Hato in the west and Fort Cimarron in the east. These actions in conjunction with the securing of Maddam Dam had the primary objective of isolating Panama City. They were also the largest of the major actions occurring during OPERATION JUST CAUSE. The airborne assault was also the largest airborne operation conducted by US forces since World War II. This large-scale shaping operation demonstrates that shaping operations are critical to mission success and can be more resource intensive than the actual operations that achieve domination.

Dominate

US Army forces achieved domination in OPERATION JUST CAUSE by establishing unchallenged military control over Panama City and eliminating Noriega's capability to challenge that control. Toward this end, the operation attacked two decisive points. The first was the assault on the PDF headquarters located in Panama City: the *Comandancia*. The second was the operation undertaken to locate and seize Noriega himself.

Three battalions of task force (TF) Bayonet (5-87th Infantry, 1-508 Infantry [Airborne], and 4-6th Infantry [Mechanized]) executed the attack on the *Comandancia* and Fort Amador. They were also tasked to protect the American Embassy in downtown Panama City. To execute these missions, they moved from various staging areas located throughout the city to their assigned objectives using air assault, mounted, and dismounted approaches. The ground movement through the city proved to be the most difficult and hazardous part of the mission due to the vulnerability of the troops in their armored personnel carriers and trucks. The dismounted movement was slower than the mounted movement but allowed the soldiers greater cover and concealment.

The strongest opposition to TF Bayonet occurred at the *Comandancia*. Elements of three PDF companies and two public order companies held out for three hours. The troops moving to *Comandancia* were subject to a large volume of sniper fire, and in the assault, unidentified indirect fire caused significant casualties among the mechanized forces. TF Bayonet forces, supported by airborne armored reconnaissance vehicles and Hellfire missiles from Apache helicopters, captured the *Comandancia*. Commanders noted in particular the precision of the supporting fires from attack helicopters. The assault by fire from supporting AC-130 gunships destroyed most of the reinforced *Comandancia* building.

Simultaneously, SOF attacked several targets where Noriega might be located. These initial attacks were unsuccessful. However, many subsequent actions neutralized Noriega's influence and eventually resulted in his apprehension on 3 January 1990. These actions included the well-organized and relentless manhunt conducted by SOF units, the isolation of Panama City itself, population control efforts, sophisticated IO, and cooperation with other US agencies.

Transition

OPERATION JUST CAUSE demonstrated the vital need for a thought-out plan that adequately addresses the transition from combat to noncombat *before* commanders initiate operations. Normally in complex UO, commanders cannot leave the details of transition until after the operation has begun without unacceptable risk to overall mission accomplishment. The follow-on stability operation, OPERATION BLIND LOGIC (later renamed OPERATION PROMOTE LIBERTY) began 24 hours after the initial assault and thus both operations were occurring simultaneously. This simultaneity of different types of operations is typical in major operations conducted in a large urban area. The stability operations involved more time than the combat operation and continued well after the close of OPERATION JUST CAUSE and after most of the major combat units had redeployed. It involved significant resources without the same level of risk to US forces as the combat operations.

Civil affairs (CA) were a dominant part of the transition from combat to stability operations. The 96th Civil Affairs battalion was central to this operation. CA forces established a civil police force, emergency food distribution, property protection, production and distribution of newspapers, cleanup of the city, and building support for a new civil government. Most tasks were coordinated through Army CA forces and executed by other Army forces under the supervision of CA.

IO were also a major aspect of affecting a stable transition and successful post-combat operations. These operations built support for the US operation among the population. They emphasized that the US conflict was with Noriega and not the Panamanian people and that the US forces would depart as soon as a new Panamanian government could take over.

Other US agencies played critical roles in stability operations in Panama. The US Drug Enforcement Agency and Justice Department were important to the negotiations that led to Noriega's capture. The US State Department helped to negotiate for Noriega and develop military policies and plans during the stability operation. The American Embassy advised commanders regarding the large diplomatic community that existed in Panama City.

FUNDAMENTALS OF URBAN OPERATIONS

5-24. UO often differ from one operation to the next. However, some fundamentals apply to UO regardless of the mission, geographical location, or level of command. Some of these fundamentals are not exclusive to urban environments. They are particularly relevant to an environment dominated by man-made structures and a dense noncombatant population (see Figure 5-4). Appendix A provides an historical example of how these fundamentals apply to an actual conflict situation.

PERFORM FOCUSED INFORMATION OPERATIONS

5-25. Information operations aimed at influencing non-Army sources of information are critical in UO. Because of the density of noncombatants and information sources, the media, the public, allies, coalition partners, neutral nations, and strategic leadership will likely scrutinize how Army forces participate in UO. The proliferation of cell phones, Internet capability, and media outlets ensure close observation of the activities of Army forces. With information sources rapidly expanding, public information of Army operations will be available faster than the internal military INFOSYS can process it. Army forces should aggressively integrate IO into every facet and at all levels of the operation to prevent negative impacts. Under media scrutiny, the actions of one soldier may have significant strategic implications. IO aim to make the information accurate; placed in the proper context of the Army's mission; and available to all interested parties: the public, the media, and other agencies.

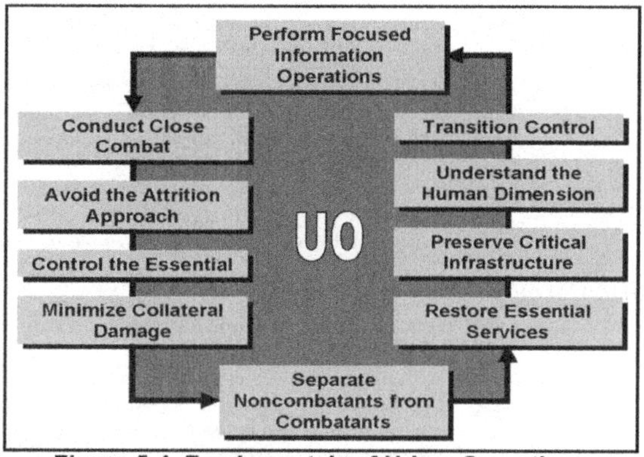

Figure 5-4. Fundamentals of Urban Operations

CONDUCT CLOSE COMBAT

5-26. Close combat is required in all offensive and defensive UO. This core capability is also present and visible in urban stability operations and may be required in urban support operations. Close combat in any urban operation is resource intensive, requires properly trained and equipped forces, and has the potential for high casualties. However, the ability to close with and destroy enemy forces as a combined arms team remains essential. This ability allows Army forces to morally dominate a threat, destroy his means to resist, and terminate urban conflicts on the Army commander's terms. Therefore, *nothing in this manual should lead commanders to compromise this decisive capability*.

AVOID THE ATTRITION APPROACH

5-27. Previous Army doctrine was inclined towards a systematic linear approach to urban combat. This approach emphasized standoff weapons and firepower. Army force structure does not support this approach towards UO. It can result in significant collateral damage, a lengthy operation, and an inconsistency with the political situation and strategic objectives. Enemy forces that defend urban areas want Army forces to adopt this approach because of the likely costs in resources. Commanders should only consider this approach to urban combat as an exception and justified by unique circumstances.

CONTROL THE ESSENTIAL

5-28. Many modern urban areas are too large to be completely occupied or even effectively controlled without an enormous force. Therefore, Army forces focus their efforts on controlling only the essentials to mission accomplishment. At a minimum, this requires control of key terrain. Key terrain is terrain whose possession or control provides a marked advantage to one side or another. In the urban environment, commanders determine key terrain based on its functional, political, economic, or social significance. A power station or a church may be key terrain.

5-29. All principles of war can apply to UO. The principle of mass and the principle of economy of force (in addition to the principle of unity of command discussed later in this chapter) are particularly important in guiding UO and providing mission focus. Army forces mass combat power only to control those requirements essential for mission success. This permits conservation of combat power. It also implies economy of force and associated risk in those areas where Army forces *choose* not to exercise control.

MINIMIZE COLLATERAL DAMAGE

5-30. Forces should use precision fires, IO, and nonlethal tactical systems consistent with mission accomplishment while decreasing the potential for collateral damage. Commanders develop unique rules of engagement (ROE) for each urban operation and provide necessary firepower constraints. IO and nonlethal systems may compensate for some restrictions, especially in stability operations and support operations. Commanders continually assess the short- and long-term effects of firepower on the population, infrastructure, subsequent missions, and national and strategic objectives.

SEPARATE NONCOMBATANTS FROM COMBATANTS

5-31. Promptly separating noncombatants from combatants (psychologically and physically) may make the operation more efficient and diminish some of the threat's asymmetrical advantages. This separation also may reduce restrictions on the use of firepower, enhance force protection, and strip the threat from its popular support base. This important task becomes more difficult when the threat is an unconventional force that can mix with civilians.

5-32. In recent operations, threats have sought to integrate their military capabilities as closely as possible into the civilian population and infrastructure. In these conditions, commanders increase their efforts to discriminate between the two. Soldiers managing violence in this setting require the highest level of individual and organizational discipline and judgment. The training, effort, and command emphasis in this area is as important as fully successful results. Such efforts strongly impact national and international perceptions of the operation.

RESTORE ESSENTIAL SERVICES

5-33. Army forces plan to restore essential services that may fail to function before or during an operation. Essential services include power, food, water, sewage, medical care, and law enforcement. When planning for and conducting Army UO, units can use nonlethal and less destructive munitions and capabilities to keep potentially vital infrastructure intact. Initially, Army forces may be the only force able to restore or provide essential services. Failure to do so can result in serious health problems for the civilians, which can affect the health of Army forces and negatively impact overall mission success. Army forces transfer responsibility for providing essential services to other agencies, NGOs, or the local government as quickly as possible.

PRESERVE CRITICAL INFRASTRUCTURE

5-34. Commanders analyze the urban area to identify critical infrastructure. They attempt to preserve the critical elements for postcombat sustainment operations, stability operations, support operations, or the health and well-being of the indigenous population. Urban areas remain in the AO after combat operations have ceased. Postcombat UO are unavoidable. Different from simply avoiding collateral damage, Army forces may have to initiate actions to prevent an enemy or a hostile civilian group from removing or destroying critical infrastructure. Such infrastructure may include cultural resources such as religious and historical places. In some cases, preserving the infrastructure may be the assigned objective of the urban operation.

UNDERSTAND THE HUMAN DIMENSION

5-35. Commanders carefully consider and manage the perceptions, allegiance, and morale of the civilians. Their assessment of the environment needs to accurately identify the attitudes of the people toward Army forces. Operational guidance to subordinates—including ROE, protection, logistics operations, and fraternization—is based on this assessment. Commanders expect and consider the demographic variance in the attitudes of an urban population. They cannot inadvertently apply Western cultural norms to a non-Western urban population. Commanders can only make assessments based on understanding and appreciating the local culture.

5-36. Sound policies, proper discipline, and adequate consideration for local culture will positively affect the attitudes of the population toward Army forces. Additionally, well-conceived and executed IO will enhance the position of Army forces relative to the urban population. Even during high-intensity urban combat, heightened awareness of and sensitivity toward the civilians can lead to a better postcombat situation than if civil considerations were unobserved or diminished in importance. An improved postcombat situation enhances transition. As the environment of conflict becomes more complex, the human dimension (and associated moral aspects) takes on greater importance and may have the greatest potential for affecting the successful outcome of UO. Therefore, the human aspect creates a discrete overall planning area.

TRANSITION CONTROL

5-37. Because UO are resource intensive, commanders plan to end them quickly, yet consistently with successful mission accomplishment. The end state of all UO transfers control of the urban area to another agency or returns it to legitimate civilian control. Quick transition releases Army resources for use elsewhere and improves the civilian morale and disposition toward Army forces. This requires the successful completion of the Army force mission and a thorough transition plan. The transition plan may include returning control of the urban area to another agency a portion at a time as conditions permit.

GENERAL EFFECTS ON OPERATIONS

5-38. Commanders understand the general effects that the environment has on the BOS. They also understand the effects that the environment has on lower-level tactics to properly plan, prepare, and execute major operations that may include UO. Otherwise, commanders may ask their subordinates to achieve effects, accomplish objectives, or adhere to a timetable that is unsupportable due to the constraints imposed by the urban environment. However, commanders do more than simply understand the impossible, rather they determine what it will take to make it possible.

BATTLEFIELD OPERATING SYSTEMS

5-39. Understanding the effects of the BOS permits the urban commander to better visualize the battlespace. See Figure 5-5. With this appreciation, he can conduct a more thorough assessment and thereby determine the most efficient means of employing Army forces. The staff can be intimately familiar with effects in their area of expertise and use that knowledge to understand the problem and develop creative and innovative solutions to achieve their commander's intent.

areas. The mass of buildings can also defuse electronic signatures. Tall buildings shield movement within urban canyons from aerial observation except from directly overhead. Urban threats may be less technology dependent and may thwart some signals intelligence efforts simply by turning off their radios and using messengers. Threat forces will likely use elements of the civilian telecommunications infrastructure for C2. These systems may include traditional landline phones, cellular telephones, and computer-to-computer or Internet data communications. Most urban telecommunications systems use buried fiber or cables or employ modern digital signaling technology. Such systems are difficult to intercept and exploit at the tactical level. These characteristics make it difficult for the intelligence system to use electronic means to determine threat dispositions and, in offensive and defensive UO, identify decisive points and centers of gravity.

5-42. **Challenging IPB Process.** The complexity of the environment also challenges the intelligence system. The intelligence system applies the IPB process to the urban environment in accordance with Army doctrine (see Appendix B). With more data points for the IPB process to identify, evaluate, and monitor, this application becomes more demanding. The human and societal aspects of the environment and the physical complexity primarily cause this difference. Relationships between aspects of the environment, built on an immense infrastructure of formal and informal systems connecting the population to the urban area, are usually less familiar to analysts. Thus, the urban environment often requires more intelligence resources to penetrate, identify, monitor, and assess than other environments.

5-43. Compounding the challenges is the relative incongruity of all urban environments. No two urban areas are alike physically, in population, or in infrastructure. Thus, experience in one urban area with a particular population and pattern of infrastructure does not readily transfer to another urban area. Any experience in UO is valuable and normally serves as a starting point for analysis, but the intelligence system cannot assume (and treat as fact) that patterns of behavior and the relationships in one urban area mirror another urban area. The opposite is as likely to hold true. The intelligence system will have to study each urban area individually to determine how it works and understand its complex relationships.

5-44. Each characteristic of the urban environment—terrain, society, and infrastructure—is dynamic and can change radically in response to UO or external influences. Civilian populations pose a special challenge to commanders conducting UO. Civilians react to, interact with, and influence to varying degrees Army forces. Commanders know and account for the potential influence these populations may have on their operations. Intelligence analysts revisit or continuously monitor the critical points looking for changes.

5-45. The actions of Army forces will affect, positively or negatively, their relationship with the urban population and, hence, mission success. NGOs may deliberately or inadvertently influence civilians. The intelligence system can monitor and predict the reactions of the civil population. Predictive analysis of a large population requires specific training and extensive cultural and regional expertise.

5-46. **Increased Importance of Human Intelligence.** The intelligence system adjusts to the degradation of its technical intelligence gathering systems by increasing emphasis on HUMINT in UO. HUMINT operations may be the primary and most productive intelligence source in UO. In urban offensive and defensive operations, HUMINT gathers information from refugees, former citizens (especially previous civil administrators), civilian contractors, and military personnel who have operated in the area. Credible intelligence of this type can help meet requirements, provide more detail, and alleviate some of the need to physically penetrate the urban area with reconnaissance forces.

5-47. In urban stability operations and support operations, HUMINT identifies threats and monitors the intentions and attitudes of the population. A chief source of HUMINT is reconnaissance forces. However, the physical nature of the urban area also poses an obstacle to these intelligence assets. Chapter 4 discusses these challenges. Reliable and trustworthy HUMINT is particularly important in foreign internal defense, combatting terrorism, and support to counterdrug operations. Leaders organize intelligence resources appropriately, and learn and apply valuable techniques, such as pattern and link analysis (see FM 34-7).

5-48. NGOs can be extremely beneficial sources of credible information about the urban environment. During the 1999 fighting in Kosovo, for example, the Red Cross provided the most accurate figures regarding the number of Kosovar refugees, helping US and other coalition forces to estimate the appropriate level of support required to handle their needs. NGOs may also have—

canalization, compartmentalization, and increased vulnerability. However, tactics and techniques equip Army forces to overcome these challenges and maneuver successfully. One tactic, effective combined arms task organization, includes an increased dismounted maneuver capability, continuous operations, and technological enhancements.

5-50. **Canalization and Compartmentalization.** The urban terrain will often canalize and compartment forces maneuvering through it. Buildings pose obstacles to both mounted and dismounted movement, forcing units to be canalized along streets. The buildings also block movement between streets, thus compartmenting units. Changing directions, repositioning committed forces, reinforcing forces in contact, bypassing threats, and maneuvering to the threat flank become extremely difficult. Units often breach obstacles to help solve this problem. Using helicopters to quickly move forces, both forward into contact and to rear areas as part of repositioning, also permits Army forces to overcome some terrain constraints.

5-51. The canalized and compartmented effects can slow maneuver considerably. However, slowed maneuver also results from the physically demanding soldier tasks required in an urban environment. Soldiers operate dismounted across rubble and hard surfaces. Operating in three dimensions, they constantly move up the intrasurface areas of building interiors and down into basements, cellars, and other subsurface areas. They breach many obstacles and use upper-body strength, ropes, and ladders to scale heights. The inability to see into the next room, floor, or building magnifies stress. The resulting fatigue slows the overall rate of Army force maneuver.

5-52. **Increased Vulnerability.** The urban environment increases the vulnerability of Army forces executing maneuver in offensive, defensive, stability, and support operations. Both the physical terrain and the urban population provide threat cover and concealment. Air maneuver is vulnerable for many of the same reasons. In offensive or defensive operations, enemy forces can remain undetected in buildings and in position to ambush Army forces. Forces clear buildings along maneuver routes prior to mounted movement along those axes. Failure to clear routes (and effectively mark cleared portions) exposes mounted movement to ambush at close range. Movement back across streets and obstacles may be difficult particularly if the element of surprise was essential in the initial crossing or breach. The same buildings also provide cover and concealment to enemy air defense capabilities, particularly man-portable air defense systems. In all operations, but especially stability operations and support operations, civilians can conceal threat elements. The threat can then initiate offensive operations against Army forces from close range and where ROE will hamper applying combat power. Thus, maneuver through a dense population can be a high-risk operation.

5-53. **Combined Arms Task Organization.** Effective combined arms task organization ensures that forces are task organized with infantry—the essential building block for all organizations conducting UO. Infantry protects mounted elements as the combined arms unit maneuvers through the urban area. The infantry destroys the enemy in buildings and bunkers where they cannot be defeated by mounted forces. Combined arms also ensure that engineers support dismounted maneuver by assisting in covered and concealed maneuver through buildings and off exposed streets. Field artillery aids in dismounted and mounted (to include air) maneuver by suppressing known and suspected enemy positions with precision fires. Armored elements protect soldiers from small arms fire and destroy or suppress enemy positions with direct fire. Artillery may also be used in this direct fire role. Armored forces and attack helicopters also can facilitate maneuver through shock action that can have a psychological effect, particularly against less well-trained threats and, in discrete instances, hostile crowds.

5-54. A major difference of UO combined arms is in proportion and organization. UO require an increased proportion of dismounted infantry and engineer capabilities. Armor is not required in the same high numbers. The level at which combined arms operations occur is also lower. Commonly, company level will require true combined arms capability and may include engineers, military intelligence, reconnaissance, and artillery. Combined arms teams can then form at platoon and squad levels. Because of this, larger units such as divisions will need more CA, military intelligence, and engineers than those included in the typical division structure, or as habitually attached for combat in more open terrain.

5-55. **Continuous Operations and Technology Enhancements.** Two other means to improve Army forces' ability to maneuver in urban terrain is through continuous operations and the leveraging of technology, such as the Army's night operations capability. Historically, urban battles have been fought primarily during daylight because of technological limitations and fatigue. By utilizing night vision technologies, accurate situational understanding, COP, training, and rotated units, Army forces can defeat threats who use the same soldiers in day and night operations and who are less well-equipped and adept at night operations. Night operations are also a means of mitigating the air defense threat against air maneuver. Continuous operations through night maneuver with fresh forces are challenging, but it can overcome many advantages that a stationary force has against maneuver in the urban environment. Commanders also consider that streetlights, fires, and background illumination (as well as dark building interiors without ambient light) may limit the effectiveness of night vision devices and make thermal imagery identification difficult.

Fire Support

5-56. The fire support system includes the collective and coordinated use of several means to attack targets in the urban area (see Appendix D for joint capabilities). These means include target acquisition data, indirect fire weapons, rotary- and fixed-wing aircraft, offensive IO, and other lethal and nonlethal means. The urban environment affects these components of the fire support system and their employment.

5-57. **Target Acquisition.** Target acquisition in an urban environment faces several challenges. First, forces have difficulty penetrating the urban environment's increased cover and concealment using sensors and reconnaissance. Acquiring targeting information and tracking targets throughout the depth of the urban area may prove challenging. Moving personnel or vehicular targets are normally easiest to acquire. However, the cover and concealment provided by urban terrain gives moving targets short exposure times requiring firing systems to act rapidly on targeting data. Targeting of opposing indirect fire units by acquisition radar works more effectively in urban terrain because of the necessary high angles of indirect fire. The urban environment presents similar difficulties for battle damage assessment.

5-58. Targeting challenges are met by innovatively integrating reconnaissance capabilities. These capabilities include SOF, long-range reconnaissance units, cavalry, unmanned aerial vehicles, and aerial observers as well as the standard reconnaissance assets of a division. More artillery systems may need to be used to ensure the responsiveness (rather than the weight) of fires. Positioning numerous artillery systems reduces the dead space (as discussed below) and permits units to establish more direct sensor-to-shooter links.

5-59. **Urban Effects on Fire Support Systems.** Both the physical and human components of the urban area affect how units use fire support weapon systems (see Figure 5-7). The physical aspects of the urban environment, such as the heights and concentration of buildings, cause significant *masking and dead space*. Buildings that stand three or more stories tall hinder close indirect fire support. Tall buildings can potentially mask several blocks of area along the gun-target line of artillery. For low-angle artillery fire, dead space is about five times the height of the building behind which the target sits. The *potential for collateral damage* to adjacent buildings may also prevent engagement with artillery. Such damage might cause noncombatant and friendly troop casualties and unintentional rubbling. Commanders can offset these effects by carefully placing artillery positions, repositioning artillery as targets change, and using mortars. Mortars have a steep angle of fall and short minimum ranges as a high-angle alternative to field artillery fire. In comparison to artillery, dead space for mortar fire is only about one-half the height of the building. Collateral damage concerns may also cause commanders to restrict attacks to certain times of day, give warning prior to an attack so that noncombatants can evacuate the area, or even abort an attack unless precision effects can be achieved.

at least 500 meters to reliably arm and stabilize on the intended target. Thus, attack helicopters firing from longer ranges actually improve the probability of a hit. Heavy smoke and dust rising from urban fires and explosions may hinder target identification, laser designation, and guidance for rotary- and fixed-winged aircraft. The close proximity of friendly units and noncombatants requires units to agree on, thoroughly disseminate, and rehearse clear techniques and procedures for marking target and friendly locations.

5-61. The urban environment also affects the *type and number of indirect fire weapon systems* employed. Commanders may prefer high-angle fire because of its ability to fire in close proximity to friendly occupied buildings. Tactically, commanders may consider reinforcing units in UO with mortar platoons from reserve units. This will increase the number of systems available to support maneuver units. Multiple Launch Rocket Systems (MLRSs) may be of limited use in urban areas due to their exceptional destructive capabilities and the potential for collateral damage. However, commanders may use MLRSs to isolate the urban area from outside influence. Commanders may also employ field artillery systems as independent sections, particularly self-propelled systems, in the direct-fire role; decreasing volume and increasing precision of artillery fire helps minimize collateral damage. While discretely applying the effects of high-explosive and concrete-piercing munitions, these self-propelled systems take advantage of the mobility and limited protection of their armored vehicles.

5-62. The urban area may affect the *positioning* of artillery. Sufficient space may not exist to place battery or platoon positions with the proper unmasked gun line. This may mandate moving and positioning artillery in sections while still massing fires on specific targets. Commanders protect artillery systems, particularly when organized into small sections. Threats to artillery include raids and snipers. Therefore, firing units will have to place increased emphasis on securing their positions.

5-63. The *mix of munitions* used by indirect fire systems will change somewhat in urban areas. Units will likely request more precision-guided munitions (PGM) for artillery systems to target small enemy positions, such as snipers or machine guns, while limiting collateral damage. Only conventional tube artillery, not mortars, has this capability. However, large expanses of polished, flat reflective surfaces common in urban areas may degrade laser designation for these munitions (as well as attack helicopter PGM). The vertical nature amplifies the geometrical constraints of many precision munitions. Remote designators need to be close enough to accurately designate but far enough away not to be acquired by the PGM during its flight path.

5-64. The urban environment also affects the use of nonprecision munitions. Building height may cause variable time fuses to arm prematurely. Tall buildings may also mask the effects of illumination rounds. Units may choose not to use dual-purpose conventional munitions if—

5-65. Depending on the building construction, commanders may prohibit or limit illumination, smoke, and other munitions because of fire hazards. (Of course, in particular instances, they may specifically use them for that effect.) Structure fires in an urban area are difficult to control and may affect friendly units. Conventional high-explosive munitions may work best against concrete, steel, stone, and other reinforced structures. When not used in the direct-fire role, a greater mass of indirect fire is often required to achieve desired effects. Commanders balance firepower and collateral damage since the rubbling caused by massive indirect fires may adversely affect a unit's ability to maneuver and provide a threat with additional cover and concealment.

5-66. Nonlethal weapons can help commanders maintain the desired balance of force protection, mission accomplishment, and safety of noncombatants by expanding the number of options available when deadly force may be problematic. As additional nonlethal capabilities are developed, they are routinely considered for their applicability to UO. In determining their use and employment, commanders consider—

5-68. **Rotary- and Fixed-Winged Aircraft.** Enemy rotary-wing aircraft can be used in various roles to include air assault, fire support, and CSS. Some threats may use unmanned aerial vehicles to obtain intelligence and target acquisition data on friendly forces. Increased air mobility limitations and targeting difficulties may cause enemy fixed-wing aircraft to target key logistics, C2 nodes, and troop concentrations outside the urban area, simultaneously attacking key infrastructure both in and out of the urban area.

5-69. **Increased Missile Threat.** The intermediate range missile capability of potential threats has increased to be the most likely air threat to an urban area. Urban areas, particularly friendly or allied, make the most attractive targets because of the sometimes-limited accuracy of these systems. By firing missiles at an urban area, a threat seeks three possible objectives:

trained and equipped for UO, can turn these obstacles into an advantage by breaching them with "mouse holes" made by explosives, sledgehammers, bulldozers or armored vehicles, or high-strength (diamond or carbide-tipped) cutting devices. These breaches permit dismounted movement through buildings under both cover and concealment.

5-74. Engineers are also trained and equipped to facilitate mounted mobility in the urban environment. Buildings are essentially unbreachable obstacles that restrict mounted movement to the compartmented and canalized streets. Threats can block streets with roadblocks ranging from sophisticated log and concrete cribs reinforced with antitank and antipersonnel mines to expedient cars, buses, and trucks. Engineers breach these obstacles to maintain the coherence of the combined arms team (mounted and dismounted). Engineers are forward, often task organized down to platoon level, and have the expertise and equipment to rapidly reduce point obstacles. It even may be necessary that every armored vehicle (or section of two vehicles) be task organized with an associated engineer squad and combat engineer vehicle.

5-75. In all UO, mobility operations may allow civilian traffic and commerce to resume, letting the urban area return to some semblance of normalcy (often a critical objective). In stability operations, mobility often focuses on keeping lines of communications open and reducing the threat of mines to soldiers and civilians. In support operations, mobility may focus on removing storm debris or reducing obstacles caused by destroyed property.

5-76. **Countermobility.** Countermobility capabilities in urban terrain are essential in all UO, not just defensive. In defensive operations, commanders use countermobility capability to control *where* the enemy moves in the urban area. Repositioning defensive forces in the urban area can be difficult and obstacles are essential to limiting the enemy's maneuver options. During offensive operations, countermobility protects exposed flanks and air assaulting forces from counterattack. In stability operations, countermobility operations may take the form of constructing barriers to assist in populace and resources control at critical urban locations.

5-77. **Survivability.** Survivability in the urban environment is a significant force multiplier. Properly positioned Army forces can take advantage of the increased survivability afforded by the physical terrain. Even a limited engineer effort can significantly enhance the combat power of small Army forces. In stability operations, properly planned and constructed survivability positions can enable small groups of soldiers to withstand the assaults of large mobs, sniping, and indirect fire. These survivability positions are often critically essential to minimizing casualties during long-term stability operations.

5-78. While executing MTW combat operations, in particular defensive operations, well planned and resourced engineer efforts can enhance the survivability characteristics of the urban area. These efforts, though still requiring significant time and materiel, can establish defensive strong points more quickly and with greater protection than can be done in more open terrain. Skillfully integrating the strong point into the urban defense greatly increases the overall effectiveness of the defense disproportionately to the number of forces actually occupying the strong point (see Chapter 7).

5-79. The Army's urban survivability operations can become complex if the Army is tasked to support survivability operations for civilians. Such operations can range from constructing civil defense shelters or evacuating the population to assisting the population in preparing for or reacting to the use of weapons of mass destruction. However, Army forces are not organized or equipped to support a major urban area's requirements as well as its own mission needs. Normally, Army forces can render this type of support only as a focused mission using a unique, specially equipped task organization.

Combat Service Support

5-80. Combat service support incorporates technical specialties and functional activities, to include maximizing available host-nation infrastructure and contracted logistics support. It provides the physical means with which forces operate. CSS operations relate to UO in two ways. The first is CSS operations conducted to support units conducting UO. The second way is conducting CSS operations from locations positioned in an urban area.

5-81. Commanders using CSS to support the full range of Army operations across the spectrum of conflict understand diverse CSS requirements of units conducting UO. They also understand how the environment (to include the population) can impact CSS support. These requirements range from minimal to extensive, requiring Army forces to provide or coordinate all life support essentials to a large urban population.

5-82. CSS commanders and staffs consider and plan for Army sustaining operations that are based in a major urban area. These operations are located in major urban areas to exploit air- and seaports, maintenance and storage facilities, transportation networks, host-nation contracting opportunities, and labor support. These operations are also UO. The CSS commander gains additional factors to consider from basing the CSS operation in an urban environment. See Chapter 9 for a detailed discussion of urban CSS.

Command and Control

> *Fighting in a city is much more involved than fighting in the field. Here the "big chiefs" have practically no influence on the officers and squad leaders commanding the units and subunits.*
>
> Soviet General Vasili Chuikov
> during the 1942-43 Battle for Stalingrad

5-83. **Command and control** is the exercise of authority and direction by a properly designated commander over forces made available in the accomplishment of the missions. He exercises authority and direction through a command and control system (FM 6-0). The urban environment influences both components of command and control: the commander and the C2 system (which includes INFOSYS). The leader's ability to physically see the battlefield, his interaction with the human component of the environment, his ability to effectively execute the targeting process, and his intellectual flexibility in the face of change all impact the mission. The C2 system faces difficulties placed on the tactical Internet and system hardware by the urban environment, by the increased volume of information, and by requirements to support the dynamic decisionmaking necessary to execute successful UO.

5-84. **Unity of Command.** Although severely challenged, the principle of unity of command remains essential to UO. However, the number of tasks and the size of the urban area often require that Army forces operate noncontiguously. Noncontiguous operations stress the C2 system and challenge the commander's ability to unify the actions of his subordinates, apply the full force of his combat power, and achieve success. To apply this crucial principle in an urban environment requires centralized planning, mission orders, and highly decentralized execution. The method of C2 that best supports UO is mission command (see FM 6-0). Mission command permits subordinates to be innovative and operate independently according to clear orders and intent as well as clearly articulated ROE. These orders and ROE guide subordinates to make the right decision when facing—

5-86. **Commander's Visualization.** Leaders at all levels need to see the battlefield to lead soldiers, make effective decisions, and give direction. Sensors and other surveillance and reconnaissance assets alone cannot provide all the information regarding the urban environment that commanders will need. The focus of lead elements narrows rapidly once in contact with a hostile force limiting their assessment to the local area. Therefore, tactical commanders will not be able to observe operations from long, stand-off ranges. Their personal observation remains as critical in urban areas as elsewhere and helps to preclude commanders from demanding their subordinates accomplish a task or advance at a rate inconsistent with the immediate situation. In urban offensive and defensive operations, seeing the battlefield requires that commanders move themselves and their command posts forward to positions that may be more exposed to risk. Thus, commanders modify their C2 system capabilities to make them smaller, reduce their signature, and increase their mobility. Because of the greater threat to C2, security efforts may be more intense.

5-87. In stability operations, commanders often intervene personally to reassure the urban population and faction leaders about the intentions of Army forces. To achieve results, commanders personally negotiate and intervene with various faction and community leaders. In these type operations, threats may attack leaders to gain the greatest payoff with the least expenditure of resources. Commanders carefully evaluate risk and potential benefits of such exposure. These risks however, cannot stop them from seeing the battlefield, personally intervening in situations as appropriate, and leading their soldiers.

> *I heard small-arms fire and RPG explosions and felt shrapnel hit the vehicle. . . . Land navigation at this time was impossible; every time I tried to look out, I was thrown in a different direction. . . . At this time, I was totally disoriented and had not realized we were on our own.*
>
> Captain Mark Hollis
> "Platoon Under Fire"

5-88. Commander's visualization also requires having detailed maps, other appropriate intelligence products, and INFOSYS that accurately depict the urban environment and help establish a COP. The reliability of these items is as important to planning major operations as it is to tactical-level operations. The commander of the major operation ensure that subordinate tactical-level commanders have the necessary products to achieve accurate situational understanding and dominate the urban environment as subordinate commands often lack the personnel or assets to develop these products. Frequently, satellite or aerial imagery is requested to compensate for the drastic changes that can occur due to UO, natural disasters, and outdated or imprecise maps. (Even maps developed and maintained by urban area's administrative activities may not be up-to-date. Extensive and continually expanding shantytowns, for example, may not be mapped at all. Maps may have even been purposefully distorted.)

5-89. Other critical intelligence products needed to visualize, describe, and direct UO may include overlays or gridded reference graphics (see also

Appendix B). Overlays and graphics portray important societal information or urban infrastructure, such as—

challenge will be communications. Urban structures, materials, densities, and configurations (such as urban canyons) and power constraints associated with man-portable radios significantly degrade frequency modulation (FM) communications. This causes problems at brigade-level and below where commanders rely heavily on constant FM radio contact with subordinates. Tactical communication problems might also cause an inability to maintain a COP, to give orders and guidance, to request support, or to coordinate and synchronize elements of the combined arms team. Communication problems in urban areas can prevent the achievement of information superiority and contribute directly to mission failure. In UO, allocating critical or high-value communication assets will be significant and essential to weighting the main effort.

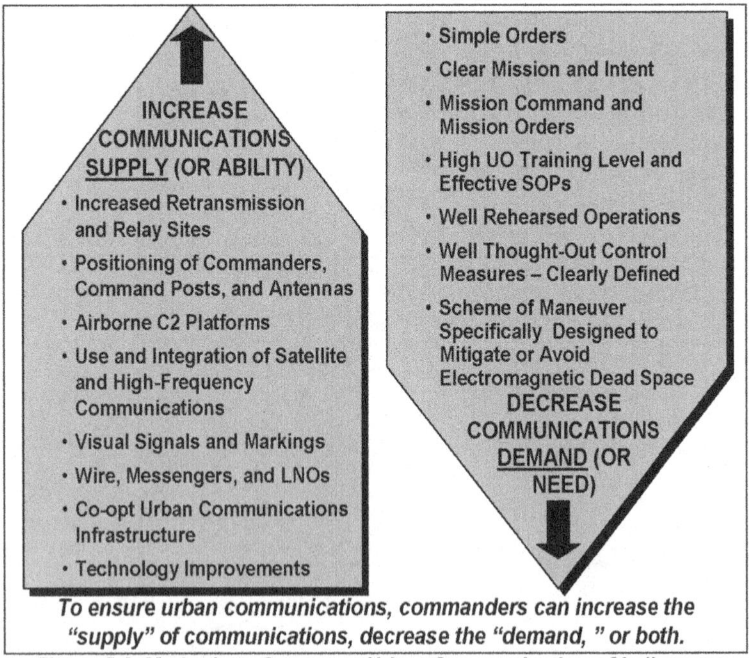

Figure 5-8. Methods to Overcome Urban Communications Challenges

5-94. In an urban environment, units and staffs properly prepare for and mitigate the communication problems in urban areas (see Figure 5-8). Adequate communications, in most cases, are ensured by—

other essential information. They coordinate with units across common boundaries. Given adequate consideration to limitations on multinational capabilities, SOPS may assist in command and control and preclude fratricide incidents resulting from loss of FM communications. However, visual signals, particularly pyrotechnics, are less effective in buildings and enclosed spaces.

5-95. In defensive, stability, or support operations, positions do not change as frequently as in offensive operations. Urban commanders then rely more on military wire (properly camouflaged amongst the civilian communications infrastructure), commercial communications, and messengers. Even in combat, some if not all of the urban area's organic communications structure remains intact for Army use. For example, every building may have one or more telephone distribution boxes that can control up to 200 individual telephone lines. Setting up wire communications using these points is relatively simple but, like all wire communications, is susceptible to wire-tapping. Cellular telephones can usually work well in urban areas; however, locating and destroying the repeater stations or the central cellular telephone system easily disables them. Consequently, the C2 system may use these alternatives to FM communications but with proper operations and physical security procedures in place.

Example of Simple Communications Innovation
Israel's Six-Day War – 1967

In the 1967 battle for Jerusalem, the Israeli Defense Force placed flags on top of cleared buildings so that aircraft providing close air support could monitor the Israeli forward line of troops. They also used a spotlight during the night to mark specific buildings as close air support targets.

5-96. Command posts above brigade-level ensure that they can communicate in an urban area without significant disruption. In stability operations and support operations, immediate and reliable communications between tactical and strategic levels may be necessary. Higher commanders anticipate that although the urban area does not significantly challenge their INFOSYS, the area may severely challenge systems at the lower tactical levels. For this reason, information flow from lower to higher may take longer. If the situation is not acceptable, the higher headquarters takes steps to mitigate it, such as increasing the number of liaison officers operating with units engaged in decisive operations. In some instances, the scheme of maneuver may be specifically designed to account for communications interference, propagation characteristics, and electromagnetic dead space. (However, this will require more time, resources, and a detailed communications IPB of the urban area.)

5-97. Finally, urban areas can overload the INFOSYS with information. UO across the spectrum of conflict and throughout the range of operations can generate large volumes of information when crises threaten. This sheer volume can easily overwhelm UO commanders and command posts. Training prepares command posts to handle this volume of information and to filter the critical from the merely informative. Staffs work hard to create products (visual or textual) that help their commanders understand the urban environment, not just present them information to know.

TACTICAL CONSIDERATIONS

5-98. Commanders and planners of major UO thoroughly understand the tactical urban battle. They especially understand the effects of the environment on men, equipment, and systems. The complexity of urban environment changes and often compresses many factors typically considered in the planning process. Figure 5-9 shows some of these compressed factors.

Figure 5-9. Compressed Tactical Factors

Time

5-99. The time available to think and act is compressed in urban combat operations. The tactical engagements that comprise battles and major UO are often quick and decisive; therefore, higher-level decisionmaking is correspondingly fast. The impact of decisions (or lack of) and the outcome of battle can occur in mere minutes. Often the amount of information and the number of decisions can overwhelm the overall ability of INFOSYS to respond. Commanders have little time to influence tactical actions with resources kept in reserve. Reserves and fire support assets are close to the point of decision so that they can respond in time to make a difference. The terrain causes C2 challenges that further inhibit commanders from responding quickly to changes in the situation. Small unit leaders receive training that emphasizes understanding the commander's intent so that they can recognize tactical opportunities and can act quickly to take advantage of them.

Distances and Density

5-100. Distances in UO are compressed to correspond to the density of threat forces and noncombatants. In open terrain, squads, platoons, and companies may be able to control or influence thousands of meters of space. In UO, large buildings can absorb the efforts of several companies or battalions. Crowds of thousands can assemble in areas of a few hundred meters requiring correspondingly large forces for control. Maximum engagement ranges, as influenced by the urban terrain, are usually closer. Units may require field artillery for direct fire at targets ranging fewer than a hundred meters. Commanders and staffs understand the telescoping nature of the battlefield, the density of threat forces, and the density of noncombatants. In addition to the actual conduct of urban tactical operations, these factors will directly affect planning, force deployment, and strength.

5-101. Time-distance considerations are especially important throughout planning cycles. Though distances may be short, the physical nature of the environment can drastically change the planning factors for unit movements. The advance of a battalion may be measured in hundreds of meters per day. Thus, all time and distance calculations that relate to sequencing of forces, synchronizing combat power and other capacities, and making decisions require reevaluation based on the urban conditions.

Combat Power

5-102. The urban terrain can also compress combat power. This terrain increases the utility and effects of some weapons and systems, increasing overall combat power. One system that dramatically demonstrates this effect is the sniper. In open terrain, snipers slightly influence operations. In UO, snipers—well concealed, positioned, and protected strategic and operational commanders to adjust ROE before the platoon has reported. Therefore, commanders at all levels know the urban environment's potential compressive effects on the levels of war. A major impact of these effects can be a lower tolerance for tactical errors and a greater need for detailed planning and precision in execution and weapons' effects (lethal and nonlethal).

Decisionmaking

5-107. The nature of the urban environment compresses the time available to make decisions and increases the number of decisions to make. This is particularly true at the lower tactical levels. Units observing an urban AO face more potential unknowns than in other situations. A large structure presents many more potential firing positions that are observed than simpler terrain. Movement in one of those windows forces the soldier or unit to quickly make a decision regarding the nature of the target—deciding whether it is a threat or a noncombatant.

Chapter 6

Urban Offensive Operations

. . . Capture Suez City "provided it does not become a Stalingrad situation."

Order to the Adan Armored Division
prior to its 1973 attack on Suez City
On the Banks of the Suez

Offensive urban operations (UO) are one of the most challenging operations that military forces can undertake. Campaigns and wars have sometimes hinged on their success or failure. Costly in resources, even when successful, they are not lightly entered into. Once engaged, they are executed rapidly and decisively. For reasons already discussed, threat forces defending in UO may gain advantages from the environment while Army force capabilities may diminish. Despite the challenges, Army forces conduct successful urban offensive operations by combining the Army's existing offensive doctrine with a thorough understanding of the environment.

PURPOSE OF URBAN OFFENSIVE OPERATIONS

6-1. Like all offensive operations, urban offensive operations are designed to impose the will of commanders on the threat. The urban offense often aims to destroy, defeat, or neutralize a threat force. However, the purpose may be to achieve some effect relating to the population or infrastructure of the urban area. Army forces may conduct offensive operations to secure a port or a communications center, to eliminate a threat to a friendly government or the urban population, or to deny the threat use of urban infrastructure. No matter the purpose, commanders use a combined arms approach for successful urban offensive operations.

CONTENTS

CHARACTERISTICS OF URBAN OFFENSIVE OPERATIONS

6-2. All offensive operations contain the characteristics of surprise, concentration, tempo, and audacity (see FM 3-0). These characteristics also apply to urban offensive operations.

SURPRISE

6-3. Army forces can achieve offensive surprise at two levels: operational and tactical. In urban offensive operations, operational surprise can be decisive. The goal is to attack the urban area before the threat expects it, from a direction he doesn't expect, or in a manner he doesn't expect. In major operations, this requires an attack against an area that appears *to the threat* to be safe from attack. Urban areas that meet this criterion are not easily accessible. Army forces launch such an attack in different ways: through a vertical assault using airborne or air assault forces, through an amphibious assault, or through a penetration followed by a rapid and deep advance. All three attacks aim to achieve surprise and to deny the threat time to prepare and establish a defense. Surprise in a major urban operation prevents a threat from falling back to occupy prepared positions in and around an urban area.

6-4. At lower tactical levels, forces achieve surprise by attacking asymmetrically. An asymmetric method attacks the threat so he cannot respond effectively. This may be achieved by using special operations forces (SOF) against a threat prepared for a conventional attack, by attacking decisively with heavy forces when the threat expects an effort by light forces or SOF, or by leveraging Army forces' extensive information operations (IO) capability. Offensive IO—primarily using IO elements of deception, electronic warfare, and operations security (OPSEC)—can help achieve surprise at all levels (see Chapter 4). Attacking at night surprises the threat and maximizes the Army forces' training, command and control (C2), and technological advantages. Attacking from unexpected or multiple directions achieves surprise by leveraging Army information systems (INFOSYS) and superior synchronization of combat power and capabilities.

CONCENTRATION

6-5. In UO, the attacking force creates a major advantage by concentrating the effects of combat power at the point and time of its choosing. The area and its compartmented effects naturally disperse and dissipate combat capability. The environment also hinders repositioning forces rapidly. Such effects can work equally against defending and attacking forces. However, in a well-prepared defense, the defender often has the advantage of interior lines. The defender can reinforce or reposition forces more quickly using covered and concealed routes (such as, sewers, tunnels, or prepared holes made in walls). Successful UO need synchronized air and ground maneuver with overwhelming effects from fires at decisive points on the urban battlefield. To achieve proper synchronization and precise effects, commanders consider the unique time and distance relationships set by the environment.

TEMPO

6-6. Tempo is the rate of military action. Commanders understand that the tempo of urban operations differs from operations in more open terrain. The complexity and the potential risk of the urban environment may invoke a cautious and methodical response on the part of commanders and their staffs. While preparing and planning urban operations, commanders conducting major operations that include urban areas strive to maintain an active tempo in offensive operations. Often, the primary purpose of the threat's urban defense is to disrupt the rapid tempo of Army offensive operations. The synchronized application of combat power and anticipation of threat reactions achieve tempo. The rapid tempo of events places Army forces in positions of advantage and helps achieve surprise. Controlling operational tempo and not allowing the different tempo of urban operations to adversely affect other operations is a challenge for commanders of major operations.

The Operational Context of Urban Operations
Brittany Ports – August to September 1944

The plan for the invasion of Normandy, France, in June 1944 was meticulously developed. The plan not only addressed the invasion itself, but also contained detailed planning for the campaign to follow. A major concern of the detailed campaign planning was logistics. To address this critical concern, and specifically the problem of ports to supply the allied armies once ashore, the pre-invasion planning called for the major ports of the French province of Brittany—Brest, Lorient, and Saint Nazaire—to be objectives of General Patton's Third Army, once it was activated.

Early August 1944, almost two months after the successful Normandy invasion, the operational situation significantly differed from that envisioned by the D-Day planners. General Montgomery's Twenty-first Army Group was still fighting in the Bocage of Normandy. In contrast, General Bradley's Twelfth Army Group had just achieved a major breakthrough at Saint Lo, secured the Cotentin Peninsula, and reached the city of Avranches. Here was a decision point. Bradley and Eisenhower had to decide whether to adhere to the original plan and turn west with Patton's forces to secure the peninsula or to take advantage of the breakout at Saint Lo and turn east to exploit the disruption of the German defenses.

Ultimately they reached a compromise. General Middleton's VIII Corps was tasked to secure the peninsula, and the bulk of Patton's Army, three Army corps, was turned northeast to exploit the operational collapse of the main German defenses. See Figure 6-1 on page 6-4.

Middleton's corps sprinted into the peninsula with the 4th and 6th Armored Divisions leading the way. However, poor communications, disagreements between commands, and contradictory orders caused the corps to hesitate before pushing the two divisions to continue to exploit toward the ports. The result: the 6th Armored Division missed an opportunity to seize Brest against light resistance by one day. The 4th Armored Division, after capturing the smaller port of Vannes, was also frustrated on the approaches to Lorient. The American reaction to the inability to rapidly seize the ports demonstrated an understanding of changing circumstances. The 6th Armored Division turned the attack at Brest to the 8th Infantry Division and then relieved the 4th Armored Division at Lorient. The 4th Armored was moved to rejoin the rest of Third Army exploiting to the east and north. Ultimately Brest fell to VIII Corps on 19 September after a 43-day siege by three infantry divisions. The victory yielded 36,000 German prisoners of war

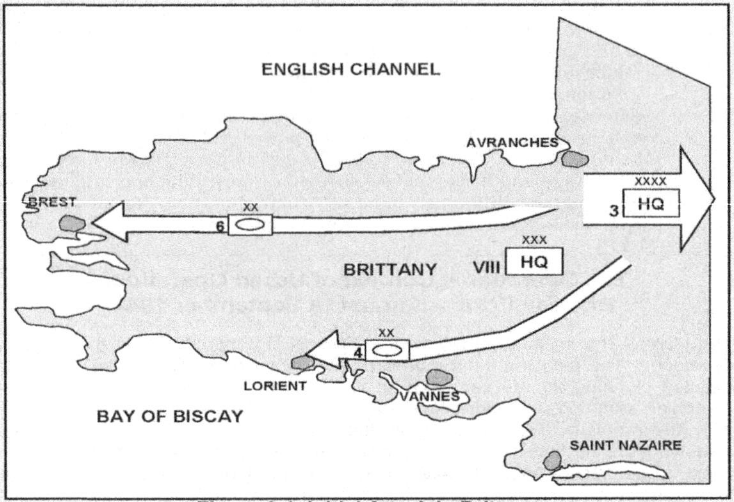

Figure 6-1. Initial Attack in Brittany

(POWs). However, the German defense and demolitions of the port left the port without an impact on the logistic situation of the allies. Brest cost the US Army almost 10,000 casualties and the commitment of significant supplies. The experience convinced commanders to surround and bypass the other major Brittany ports. Lorient and Saint Nazaire remained under German control, deep in allied territory, until the war ended ten months later (see Figure 6-2).

The operational lessons of the Brittany campaign are numerous. First, commanders are responsible to continually assess assumptions and decisions made during planning based on the changing circumstances of the battlefield. This includes the planning decision to conduct urban offensive operations. When the allies arrived at the Brittany Peninsula, the focus of the operational maneuver was no longer securing logistics facilities but exploiting the breakthrough at Saint Lo and the disintegrating the German defense. The bulk of Third Army then was turned to the north and east rather than west into the peninsula.

The Brest experience also demonstrates that the costs of urban offensive operations are continually assessed against the operational value of the objective. This lesson was applied to the cities of Lorient and Saint Nazaire. The cities were never seized from the Germans because their logistic value failed to warrant the required resources. German retention of the ports had no major adverse effect on the overall campaign.

Another lesson is that commanders cannot allow urban operations to disrupt the tempo of other offensive operations. One German goal of defending the ports was to disrupt the rapid tempo of the US exploitation. They failed to achieve this

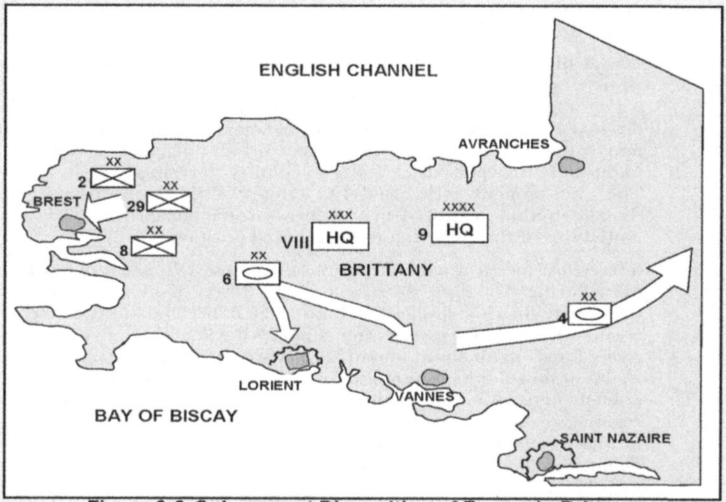

Figure 6-2. Subsequent Disposition of Forces in Brittany

goal because General Bradley continued the exploitation with the bulk of Third Army and executed the original plan with only a single corps.

Finally, commanders cannot allow emotion to color their decision to conduct or continue UO. The failure of 6th Armored Division to seize Brest rapidly caused some commanders to believe that Brest had to be captured because the prestige of the Army was committed to the battle. Costs of the continuing combat operations to seize Brest were significant. These resources might have been better committed elsewhere in the theater.

6-7. Tactical tempo is also important in urban combat. Because of the complex terrain, defending forces can rapidly occupy and defend from a position of strength. Once Army forces initiate tactical offensive operations, they cannot allow the threat to set the tempo of the operation. Instead, attacking forces seek to maintain a high tempo of operations. However, the tactical tempo of urban operations differs from operations in other terrain. Not necessarily slow, it requires a careful balance of preparation, speed, and security. In terms of unit fatigue, resource consumption, and contact with the threat, the tempo of most urban offensive operations may be rated as very high. On the other hand, in distances traveled and time consumed to achieve objectives, the tempo of many urban offensive operations might be rated as slow. The urban battlefield's density concentrates activity and consumes resources in a relatively small area. The lack of terrain seized or secured is not to be construed to mean a low tempo in the battle. In reality, the natural tempo of urban operations is not faster or slower than other types of operations.

merely different. A higher tempo of operations, however, can favor forces which are better led, trained, prepared, and resourced.

6-8. A high tactical tempo in urban offensive operations challenges logisticians to provide for the increased consumption of munitions and degrades soldiers' physical capabilities. Commanders anticipate these challenges and develop the means and abilities to overcome them. In the past, these challenges forced commanders to conduct urban offensives cyclically. They used night and other periods of limited visibility to resupply, rest, and refit forces. The environment influenced the tempo of their operations. This type of "battle rhythm" resulted in the forces spending each new day attacking a rested threat that was in a well-prepared position.

6-9. Army forces must maintain the tempo. Offensive operations continue even during darkness. Moreover, Army forces increase the tempo of operations at night to leverage the limited visibility capabilities, increased situational understanding, training, and INFOSYS that give an advantage to Army forces in all environments. To overcome the physical impact of the environment on soldiers, commanders retain a large reserve to rotate, continuing offensive operations at night. The force that fights in daylight becomes the reserve, rests, and conducts sustaining operations while another force fights at night. Army forces can then maintain the tempo of operations and leverage technological advantages in urban offensive combat.

6-10. Tempo in UO does not necessarily mean speed. Offensive operations balance speed, security, and adequate firepower. Commanders plan for the complex tactical environment and the requirements to secure flanks and airspace as the operation progresses. Mission orders allow subordinate units to make the most of tactical advantages and fleeting opportunities.

AUDACITY

6-11. Audacity is a simple plan of action, boldly executed. Superb execution and calculated risk exemplify it. In an urban attack, a thorough assessment of the physical terrain can mitigate risk. The terrain's complexity can be studied to reveal advantages to the attacker. Audacity can also be embodied in an operation by inventively integrating the direct action tasks of SOF throughout the operation. Combining SOF actions with conventional attacks can asymmetrically unhinge a defensive plan.

URBAN OFFENSIVE OPERATIONS AND BATTLEFIELD ORGANIZATION

6-12. Urban offensive operations, like all operations, are framed in the overall doctrinal framework of sustaining, shaping, and decisive operations. Each operation is essential to the success of an urban offensive, and usually two or more of these operations occur simultaneously. Sustaining operations in urban offensive operations ensure freedom of action. They occur throughout the area of operations (AO) and for the duration of the operation. Shaping operations in urban offensive operations create the conditions for decisive operations. In UO, much of the shaping effort focuses on isolation, which is critical in both major operations and tactical battles and engagements. Decisive operations are attacks that conclusively determine the outcome of UO.

These attacks strike at a series of decisive points and directly lead to neutralizing the threat's center of gravity.

SUSTAINING OPERATIONS

6-13. Commanders conducting urban offensive operations ensure security of the sustaining operation and bases; in many situations, sustaining operations may be the greatest vulnerability of the attacking force. Those supporting an urban offensive are tailored to the urban environment and are well forward. Ideally, the supporting forces closely follow the combat forces and move within or just outside the urban area as soon as they secure an area. Operating in the urban area during offensive operations allows the sustaining operation to take advantage of the defensive attributes of the environment for security purposes.

6-14. Counterattacks against sustaining operations may take the form of special operations activities aimed at lines of communications (LOCs) leading to or within the urban area. Choke points—such as bridges, tunnels, and mountain passes—are vulnerable to these attacks and may require combat forces to protect them. Threat forces attack the LOC to blunt the Army's combat power advantage in the urban area.

6-15. Attacks against the LOC into the urban area may also attempt to isolate the attacking Army forces from its sustainment base. Isolated forces in an urban area are greatly disadvantaged. Commanders plan and aggressively execute strong measures to protect their LOC, even if it requires reduced combat power to execute their offensive operation.

6-16. Sustaining operations anticipate the volume and unique logistics requirements of urban operations. Specialized individual equipment—such as grappling hooks, ladders, and pads—is identified and provided to troops in quantity before they are needed. Forces stockpile and distribute their attacking units' special munitions requirements including small arms, explosives, and grenades of all types, precision artillery munitions, and mortar ammunition. Forces also supply transport to move the resources rapidly forward, both to and through the urban environment. Sustaining operations cannot rely on "operational pauses" to execute their tasks. Commanders plan to continuously supply resources and capabilities to the most forward combatants as offensive operations advance.

6-17. Sustaining operations also anticipate the growth of sustainment requirements as Army forces secure and take responsibility for large portions of the urban area. The success of Army urban offensive operations will often uncover the civil population in former threat occupied areas. It may attract the civil population from sections of the urban area where the Army is not operating to areas occupied by Army forces. Rural populations may migrate to the urban area as the result of successful Army offensive operations.

6-18. Army forces may be required to take initial responsibility to provide for the urban population. This consideration is integrated into logistics planning and organization from the start of the planning process. To be successful and efficient in such a situation, logistics planning includes Army civil affairs (CA) specialists and local government representatives. It also integrates and consults with the international community and nongovernmental organizations (NGOs) that might augment or supplement Army logistics capabilities.

SHAPING OPERATIONS

6-19. Shaping operations that support the urban attack separate into those focused on isolating the threat and all others. Army forces isolate the threat to ensure successful urban offensive operations. Depending on the threat reaction to isolation efforts and the nature of the threat center of gravity, this task may become decisive. Other shaping operations include those common to all offensive operations and others unique to urban operations. Unique urban shaping operations may include securing a foothold in a well-fortified defensive sector, securing key infrastructure, or protecting noncombatants. Because of the nature of UO, shaping operations may consume a much larger proportion of the force than during other operations and may take place both inside and outside the urban area (see Applying the Urban Operational Framework: Panama in Chapter 5). By successfully isolating a threat force, the force needed to conduct the decisive operation may be relatively small.

DECISIVE OPERATIONS

6-20. A tactical commander fights decisive urban combat, whereas commanders conducting a larger major operation influence urban combat by setting the conditions for tactical success. Higher commanders may directly influence urban offensive operations by operational maneuver, by coordinating joint fires, by closely coordinating conventional forces, or with SOF.

6-21. Tactical urban offensive operations quickly devolve into small-unit tactics of squads, platoons, and companies seizing their objectives. The compartmented effect of the terrain and the obstacles to command and control of small units, especially once they enter close combat inside buildings or underground, often restricts the higher commander's ability to influence operations. Commanders influence the actions of subordinates by clearly identifying the center of gravity and decisive points; using mission orders (as discussed in Chapter 5); developing effective task organizations; and synchronizing their sustaining, shaping, and decisive operations.

6-22. Like all operations, successful decisive operations in UO depend on identifying the decisive points so the forces can destroy or neutralize the threat's center of gravity. Seizing a key structure or system that makes the threat's defense untenable; interdicting a key resupply route that effectively isolates the threat force from his primary source of support; or isolating the threat so that his force can no longer influence friendly activity may be more effective than his outright destruction.

6-23. Commanders select the right subordinate force for the mission and balance it with appropriate attachments. Higher commanders do not direct how to organize the small tactical combined arms teams, but they ensure that subordinates have the proper balance of forces from which to form these teams. Successful urban offensive operations require small tactical combined arms teams. Urban offensive operations require abundant infantry as the base of this force. However, successful urban combat requires a combined arms approach (of which armored and mechanized forces will be essential) adjusted for the conditions of the environment. Precision-capable artillery systems generally support urban operations better than rocket artillery.

6-24. Divisions entering urban combat may require additional resources. These resources include military intelligence support in the form of linguists, human intelligence (HUMINT) specialists, and unmanned aerial vehicles (UAVs). Engineering assets will be at a premium; the task organization of a task force executing the decisive operation may require a one-to-one ratio of engineer units to combat units. Corps and higher engineering support may be necessary to meet these requirements and to repair vital and specialized infrastructure. A tailored and dedicated corps support battalion or corps support group may need to assist in providing anticipated support to a displaced and stressed civil population. Finally, divisional CA units may require augmentation to deal with NGOs and civilian government issues.

6-25. Successfully conducting decisive operations in the urban environment requires properly synchronizing the application of all available combat power. Army forces have a major advantage in the command and control of operations. Commanders use this advantage to attack numerous decisive points simultaneously or in rapid succession. They also use it to attack each individual decisive point from as many directions and with as many different complementary capabilities as possible. Commanders completely understand urban environmental effects on the battle operating systems to envision and execute the bold and imaginative operations required. Significantly, these operations require that C2 systems account for the mitigating effects of the environment as execution occurs.

6-26. Properly synchronized actions considerably enhance the relative value of the combat power applied at the decisive points. They present to the threat more requirements than he has resources with which to respond. Synchronized IO and multiple maneuver actions paralyze the threat's decisionmaking capacity with information overload combined with attacks on his C2 systems. Additionally, well-synchronized actions limit the time the threat has to make decisions and forces him into bad decisions. In the urban environment, these effects are enhanced because C2 systems are already strained, poor decisions are harder to retrieve, and units that do not react are isolated and destroyed.

FORMS AND TYPES OF URBAN OFFENSE

6-27. Traditional forms of offensive maneuver include envelopment, turning movement, infiltration, penetration, and frontal attack. Traditional types of offensive operations are movement to contact, attack, exploitation, and pursuit. These traditional forms listed apply to urban combat. Some have greater application to an urban environment than others do. Moreover, success will belong to commanders who imaginatively combine and sequence these forms and types throughout the depth, breadth, and height of the urban battlefield. This is true at the lowest tactical level and in major operations.

FORMS OF OFFENSIVE MANEUVER

Envelopment

6-28. The envelopment is the ideal maneuver for isolating threat elements in the urban area or isolating the area itself. A deep envelopment effectively isolates the defending forces and sets the conditions for attacking the urban area from the flank or rear. Yet, enveloping an objective or threat force in the urban area is often harder since achieving speed of maneuver in the environment is so difficult (see Figure 6-3). Vertical envelopment, however, works effectively if Army fires can effectively suppress or neutralize the threat air defense.

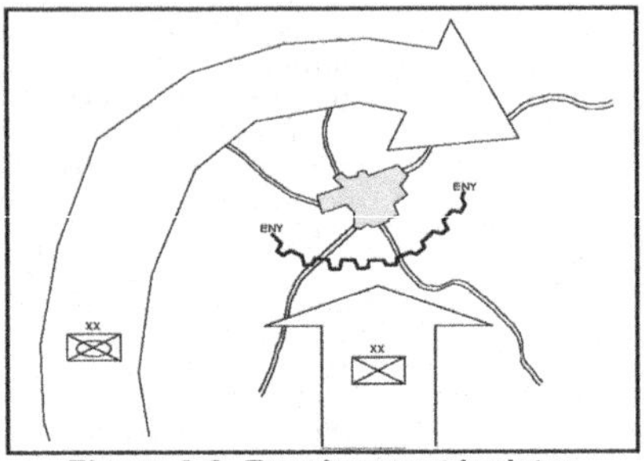

Figure 6-3. Envelopment Isolates an Urban Area

Turning Movement

6-29. Turning movements can also be extremely effective in major operations (see Figure 6-4). By controlling key LOCs into the urban area, Army forces can force the threat to abandon the urban area entirely. These movements may also force the threat to fight in the open to regain control of LOCs.

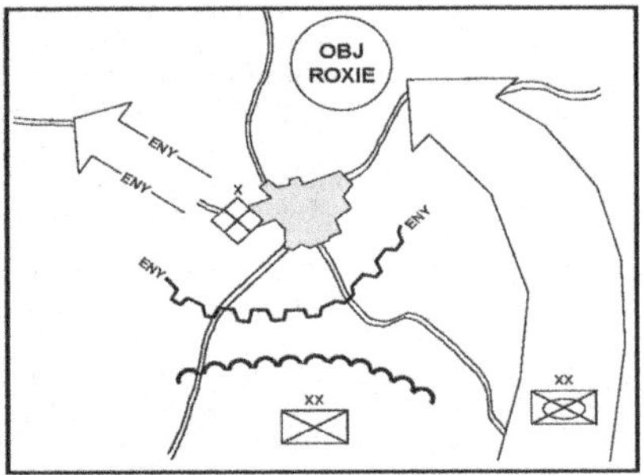

Figure 6-4. Turning Movement

Infiltration

6-30. Infiltration secures key objectives in the urban area while avoiding unnecessary combat with threat defensive forces on conditions favorable to them (see Figure 6-5). This technique seeks to avoid the threat's defense using stealthy, clandestine movement through all dimensions of the urban area to occupy positions of advantage in the threat's rear (or elsewhere). It depends on the careful selection of objectives that threaten the integrity of the threat's defense and superior COP. Well-planned and resourced deception operations may potentially play

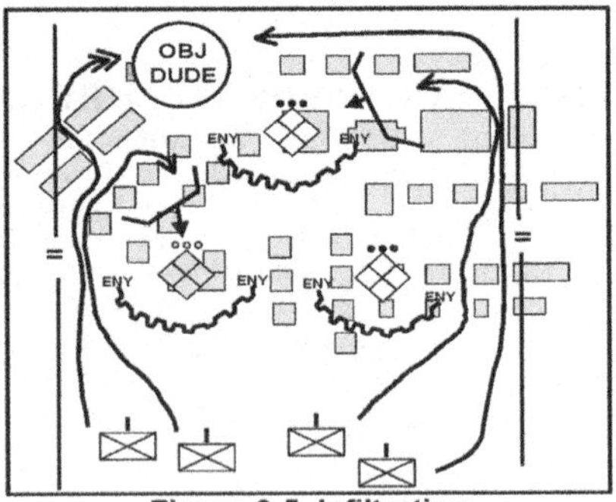

Figure 6-5. Infiltration

a critical role in masking the movement of infiltrating forces. The difficulty of infiltration attacks increases with the size and number of units involved. It is also more difficult when Army forces face a hostile civilian population. Under such circumstances, infiltration by conventional forces may be impossible. Armored forces are generally inappropriate for infiltration operations. However, they may infiltrate large urban areas if the threat is not established in strength and had insufficient time to prepare defenses.

Penetration

6-31. Penetration is the most useful form of attack against a prepared and comprehensive urban defense (see Figure 6-6). It focuses on successfully attacking a decisive point. Ideally in urban combat, multiple penetrations in all dimensions are focused at the same decisive point or on several decisive points simultaneously. In urban combat, the flanks of a penetration attack are secure, and resources are positioned to exploit the penetration once achieved.

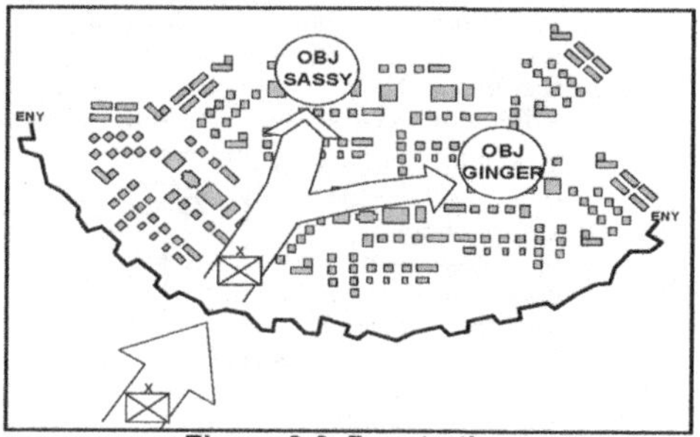

Figure 6-6. Penetration

Frontal Attack

6-32. Frontal attacks are the least favorable form of maneuver against an urban area (see Figure 6-7). They require many resources to execute properly, risk dispersing combat power into nonessential portions of the area, and risk exposing more of the force than necessary to threat fires. In urban offensive combat, forces most effectively use the frontal attack at the lowest tactical level once they set conditions to ensure that they have achieved overwhelming combat power. Then the force of the frontal attack overwhelms the threat with speed, firepower, and shock action.

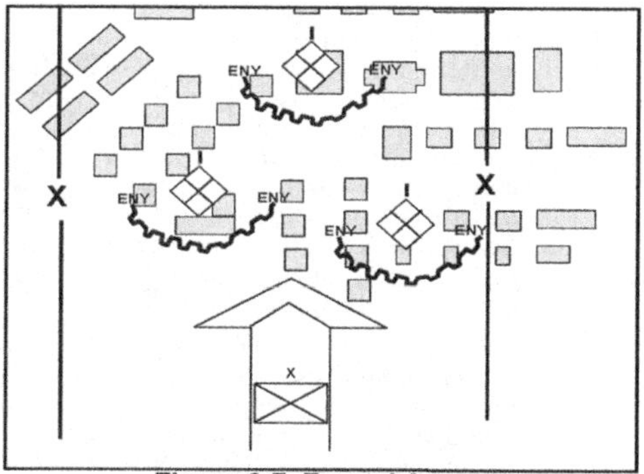

Figure 6-7. Frontal Attack

Forms of Attack in the Urban Offense
Metz – 1944

In November 1944, the US Third Army launched its final effort to take the French city of Metz from the defending Germans. This was the Army's third attempt. The first attempt had been a surprise, mounted attack. This was followed by a series of piecemeal infantry assaults on the surrounding fortresses. Finally, a deliberate effort was made to take the city in a coordinated effort by XX Corps.

The initial failures stemmed from a shortage of resources on the US side, to include fuel and units—especially infantry. This added to the ad-hoc nature of the first two efforts. The third effort, though more deliberate, was still constrained by resources. XX Corps could only muster three nearly full-strength infantry divisions to attack the German's defending with four under-strength divisions.

The third attempt to take Metz demonstrates how a corps operates with multiple divisions using various forms of attack to achieve its objective in urban offensive operations.

The opening phase of the Metz battle had attacks by the 90th and 5th Infantry Divisions to envelop the city from the north and south (see Figure 6-8). This isolated the city and ensured the garrison could not escape nor be reinforced. The garrison was under orders from Hitler not to conduct a breakout and to resist to the last man; thus, German forces strongly counterattacked both wings of the envelopment to prevent isolation.

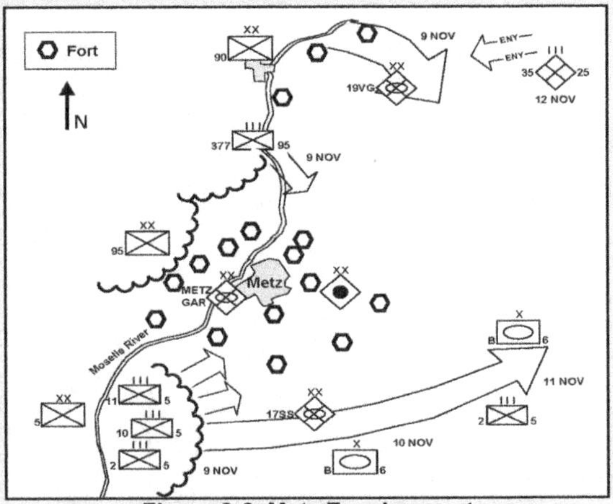

Figure 6-8. Metz Envelopment

The second phase of the operation was the penetration of the city defenses from multiple directions. The 5th Infantry Division penetrated into the city with the 10th and 11th Infantry Regiments from the south. The 95th Infantry Division penetrated into the city from the north with the 377th Infantry Regiment and TF Bacon. Simultaneously, the 95th Infantry Division infiltrated battalions through the string of fortress positions guarding the western approaches into the city, isolated, and bypassed these positions with its other two regiments.

The final reduction of the defense was a series of battalion frontal attacks, which took place against the last remaining strongholds within the city. Even in these final engagements, however, the infantry battalions isolated, bypassed, then attacked from the flanks and rear whenever possible (see Figure 6-9).

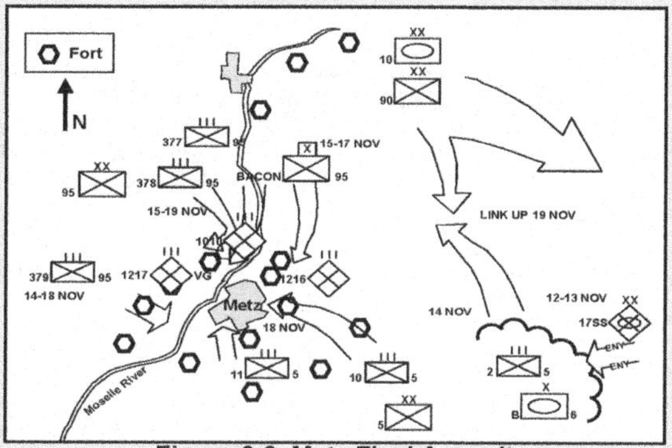

Figure 6-9. Metz Final Assault

The city was declared secured on 19 November. However, at that point more than a half-dozen of the fortresses had yet to be reduced. The 95th Infantry Division, after a four-day rest, quickly moved forward to rejoin the still rapidly advancing corps forward elements. Elements of the 5th Infantry Division remained in siege posture around the remaining strong points, the last of which surrendered on 19 December 1944 when it ran out of food. US forces made no efforts to attempt to assault these bypassed fortresses although extensive psychological operations (PSYOP) were used.

TYPES OF OFFENSIVE OPERATIONS

Movement to Contact

6-33. In an urban area where the threat situation is vague, Army forces conduct a movement to contact to establish or regain threat contact and develop the situation. Often a movement to contact in an urban area occurs as both sides try to establish their influence or control over a contested population center. The situation determines whether the movement to contact or its specific technique, the search and attack, is appropriate. A conventional force-oriented movement to contact will likely take place when friendly and threat conventional forces attempt to establish control simultaneously. Initially, neither side is defensive. The friendly force aims to quickly locate and fix the threat while establishing control of the urban area and its key infrastructure. The search and attack technique works well when a smaller threat has established a noncontiguous defense in an urban area. This operation is characterized by the friendly point defense of key infrastructure, robust reconnaissance, and rapidly concentrated combat power to fix and defeat or destroy threat resistance once located.

6-34. A meeting engagement often results from the movement to contact. It occurs when a moving force that is partially deployed for battle collides with and engages a threat at an unexpected time and place. In a meeting engagement in an urban area, the unit that reacts most quickly and decisively will likely win. Rapid and accurate decisionmaking depends heavily on understanding the nature of the urban area and its impact on operations. Thus, in a meeting engagement, commanders quickly assess the impact and role of all components of the urban environment (terrain, infrastructure, and society) on the operation. Responsive reconnaissance and situational understanding are also important. This permits accurate decisionmaking regarding where to attack, where to defend, and how to allocate resources. Situational understanding enhanced by digital INFOSYS that provide an enhanced common operational picture (COP) facilitates the rapid reaction of Army units and a synchronized response. This reaction and response allow Army forces to seize the initiative and dominate the threat.

Attack

6-35. The attack is the most common and likely offensive operation that Army forces will conduct in an urban environment. Commanders conducting major operations and commanders of large tactical units usually execute deliberate attacks. In the urban environment, units larger than battalion-size rarely conduct hasty attacks. Hasty attacks are common below company level as units use their initiative to take advantage of tactical opportunities. Larger units will conduct hasty attacks when threat defenses are disrupted or unprepared, to take advantage of an unexpected situation, and to prevent the threat from establishing or re-establishing a coherent defense.

Exploitation

6-36. Exploitation follows a successful attack to disrupt the threat in depth. Commanders of major operations consider focusing exploitation attacks on urban areas. A threat defeated in an attack will attempt to rally units, reinforce with reserves, and reorganize his defense. With its information and communications capability, transportation network, and defensive attributes, the urban area is the natural focal point to reestablish a disrupted defense. By establishing urban centers as the objectives of the exploitation, commanders deny the threat the location he needs to reestablish his defense. The exploitation focuses on the urban area as well as on the remnants of the threat. A successful exploitation to seize an urban area works efficiently because the attack preempts the defense and denies the threat the full advantages of urban terrain.

6-37. Commanders conducting exploitation acknowledge the vulnerability of their forces to counterattack and ambush in urban areas. An urban area provides ideal cover and concealment to hide threat reserves, reinforcements, or reorganized forces. Constrictions of routes into and through the urban area make exploitation forces a potentially dense target and limit maneuver options. Robust and well-coordinated reconnaissance, tactical dispersal, and use of advance guard security forces protect against this threat (see Defensive Combat Power: Suez City vignette in Chapter 7).

Pursuit

6-38. The pursuit is designed to destroy threat forces attempting to escape. It focuses on the threat and not on urban areas. When conducting a pursuit, Army forces move through undefended urban areas and bypass those in which threat forces successfully take refuge. The threat will likely attempt to use urban areas to disrupt the pursuit and permit the threat main body to escape. Commanders prevent escape by denying the threat the time to establish forces in urban areas that cannot be bypassed. The agility of Army aviation forces for attack, reconnaissance, and transportation is essential to execute a successful pursuit around and through urban areas.

URBAN OFFENSIVE CONSIDERATIONS

6-39. The urban operational framework (assess, shape, dominate, and transition) provides a structure for developing considerations unique to urban offensive operations. The considerations vary depending on the situation and scale of the operation. Some considerations applicable to major operations that include an urban area will also be considerations at the tactical level focused in the urban area. However, no set rules exist. All urban operations are unique. Issues addressed at the operational level in one situation may be addressed in a new situation only at the tactical level. Under the right circumstances, a consideration may become an operational issue, a tactical issue, or a combination of the two. The following identifies some planning and execution issues that commanders conducting major operations address.

ASSESS

6-40. The first requirement, and a continuing requirement throughout the conduct of urban operations, is the assessment of the situation. Commanders base this assessment on detailed information regarding the particular urban area. Since the threat will likely dominate or control most of the urban area during the planning phase of offensive operations, accurate assessment of the urban environment will be difficult. A comprehensive intelligence, surveillance, and reconnaissance (ISR) effort in support of a rigorous intelligence preparation of the battlefield (IPB) process overcomes this obstacle.

Integrated Intelligence, Surveillance, and Reconnaissance

6-41. The commander of a major operation that includes an urban area, unlike his subordinate commanders, can target reconnaissance deep into the AO and area of interest. This begins the application of ISR resources against the urban area that may lead to decisive ground operations. This ISR effort and the assessment it supports continue as long as the urban area remains in the AO. Commanders of major operations initially direct ISR assets on those information requirements that support determining whether or not to conduct urban offensive operations. Once decided, ISR resources shift to support the planning and execution of the operation in the urban area.

6-42. The first resources that a senior commander can use are national and strategic sensors. He requests them through the appropriate joint force commander. The commander aggressively pursues full use of these systems to begin building an initial database for analyzing the significant aspects of the terrain; key infrastructure considerations; the status and disposition of the population; and the size, type, and disposition of threat forces in the area.

6-43. Simultaneously, multiple intelligence sources contribute to the database. The sources collect, process, store, display, and disseminate the relevant information on large urban areas through open and classified resources. These information sources include—

on the essential, they may determine the decisive point to be a single building or a limited sector of an urban area. It could be an entire system within the urban infrastructure such as communications and information, or a limited subsystem of the transportation and distribution infrastructure such as a single airfield. Sometimes what is decisive in the urban area is the threat military capability, but even this large an objective, when carefully analyzed, may not require destruction of all threat forces or control of all a large urban area. Decisive points relate directly to the threat's center of gravity and to mission success. Some decisive points related to the urban threat's center of gravity may be physically located outside the urban area.

6-48. To gain specifics on threat dispositions within the urban area requires reconnaissance capability to see into the depths of the area and the intelligence capability to determine the threat's likely defensive course of action. With this information, commanders can determine decisive points and apply Army combat power discretely against them. Effective urban offensive operations require detailed situational understanding of an area of interest that extends well beyond the perimeter of the urban area.

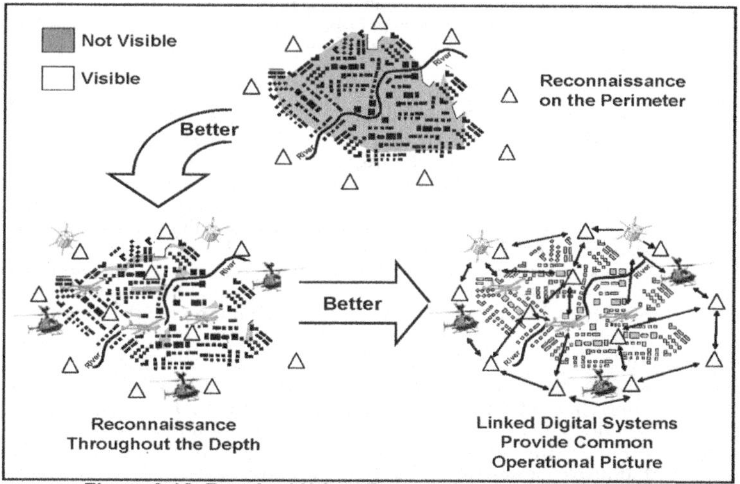

Figure 6-10. Required Urban Reconnaissance Capabilities

6-49. Commanders see throughout the depth of the urban area using several actions (see Figure 6-10). First, they evaluate sensor data and imagery. This guides targeting of special reconnaissance. Simultaneously, HUMINT is conducted using any persons who might know the urban area and threat. This includes civilians (allies, aides, neutrals, obstacles, and hostiles) and POWs. Finally, tactical conventional reconnaissance assets including reconnaissance forces, aviation, artillery radar, signals intelligence, and UAVs are directed at the urban area. All these sources and data are linked through digital

INFOSYS to provide commanders and their subordinates with improved situational understanding and a COP.

6-50. **Collateral Damage Considerations.** Commanders also assess the collateral damage risks that his operation may include. This assessment helps to initially determine the viability of a course of action. However, commanders reassess their courses of action at frequent intervals in urban offensive operations based on known information to determine if the original evaluations remain valid. This reassessment minimizes potential collateral damage from a change in mission or a change in a course of action. Many aspects of the environment can change during mission execution.

6-51. **The Environment's Effects on Courses of Action.** The urban environment's unique aspects can significantly impact the course of action chosen by Army forces and the threat. Commanders assess these effects in planning, but they also verify and monitor these effects as forces execute offensive missions. In particular, commanders will want to confirm the civilian population's locations, beliefs, and actions and to monitor any changes. They will need to validate terrain considerations and monitor how any changes based on rubble and other damage to structures. In urban terrain, dead space, cover, and concealment can only be identified physically and will change considerably as operations affect the terrain.

SHAPE

6-52. Commanders of major operations have a primary contribution to urban operations: the planning and conduct of effective shaping operations that set the conditions for subordinate tactical success. In urban operations, isolation will be a critical condition. Effective isolation will require persistent, continuous surveillance and reconnaissance, innovative use of fires and maneuver (including effective force allocation decisions), and well-established sensor-to-shooter links. These efforts—combined and synchronized with SOF's direct actions, IO that minimize noncombatant influences, and necessary shaping attacks (particularly the seizure of a foothold)—establish the conditions necessary for the subsequent offensive domination of the area.

Isolation is Essential

6-53. One key to success in the history of urban operations has been the effective isolation of the threat force (see Figure 6-11). This applies today and equally well to major urban offensive operations as it does to smaller-unit attacks. This isolation not only denies access to the urban area from outside but also contains threat forces within. In a modern metropolis or megalopolis, this can appear a daunting task. Operational isolation requires dominating all physical and electronic contact between the threat in the urban area and supporting threat forces outside the urban area. This does not necessarily require physically encircling the urban area, but it does require that Army forces be able to exert *control* over the area's entire perimeter, as well as decisive points within. For a sprawling urban area, successful isolation may require the commitment of a large amount of resources.

6-54. Successful isolation of the urban area depends as much on the nature of the threat as it does on any other factor. A conventional threat in a large urban area may be much easier to isolate than an insurgent threat in a much smaller urban area. The forces needed in the former situation may be less than those needed in the latter. The more the characteristics of the threat are conventional in nature generally the easier it will be to isolate him using standard combat methods and equipment. Isolating a more unconventional force requires many of the same techniques as used against conventional forces. It also requires a much greater ability to conduct offensive IO, to integrate CA units and civil-military operations (CMO), and to work with allies and local authorities. Fundamentally, isolating a less conventional threat puts increased emphasis on separating combatants from noncombatants.

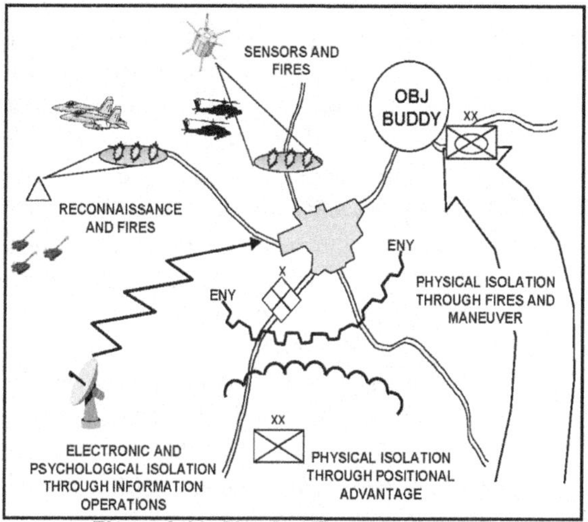

Figure 6-11. Shaping Through Isolation

6-55. **Offensive Isolation Objectives.** Isolation seeks to achieve two primary objectives with respect to defeating a threat's urban defense:

monitored by sensors or observation, are integral to any isolation technique.) First, attacking forces can pre-position themselves along avenues of approach to deny entry and exit through positional advantage. Relying primarily on this method of isolation, particularly around a large urban area with multiple avenues of approach, can be resource intensive. Instead, the pairing of fires and maneuver provides attacking commanders more flexibility and allows them to isolate several avenues of approach with fewer resources. Highly mobile attack helicopters are ideal for this purpose as long as these operations occur outside the threat-controlled portions of the urban area. In these threat-controlled areas, it is more difficult to identify, eliminate, or effectively suppress the air defense threat. The threats may have numerous man-portable air defense weapons and enhanced effects of small arms used for air defense. Therefore, the risk to using this equipment may outweigh the potential benefits. However, mobile ground units—such as an air assault (subject to the same air defense threat considerations as attack aviation), armored, or mechanized forces—can also rapidly move to attack and destroy a threat moving in or out of an urban area. Potential disadvantages of the combined, fires and maneuver, option are that the—

units will then respond to the results of surveillance depending on the circumstances.

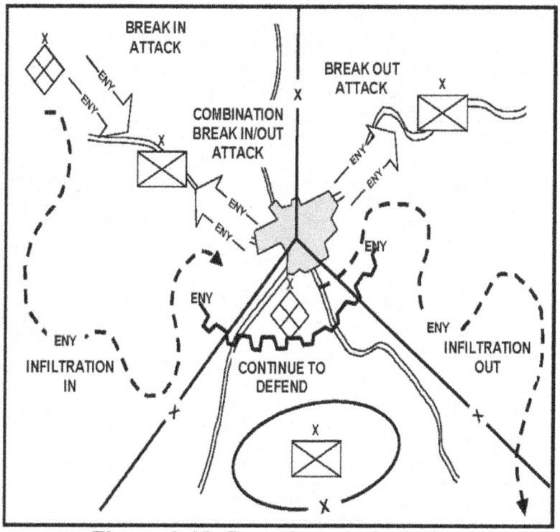

Figure 6-13. Reactions to Isolation

6-59. **Threat Reactions.** The reaction of the threat to the effects of isolation will depend on his mission, morale, force structure, and overall campaign plan. The threat may recognize isolation actions early and withdraw from the urban area before isolation is completed instead of risking destruction. On the other hand, the threat, based on a different or flawed assessment (perhaps a perception shaped by the Army force commander), may choose to—

Both ARVN and US forces swiftly reacted to the NVA attack. Within 24 hours, ARVN infantry and cavalry units counterattacked to recapture the Citadel. South of the Perfume River, the 1st Marine Division began attacking to clear the southern portion of the city (see Figure 6-14).

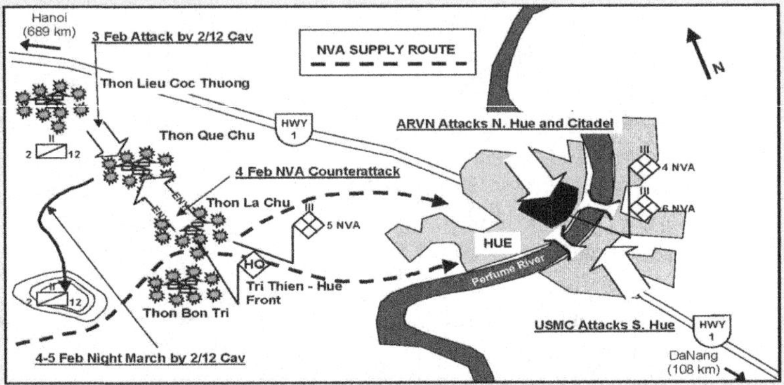

Figure 6-14. Initial Attack to Isolate Hue

On 2 February, the first Army unit was committed to Hue. The 2-12th Cavalry (an infantry battalion) was ordered to attack southeast along Highway 1 and clear enemy interdicting that route. On 3 February, 2-12th Cavalry began their attack moving along the south side of the highway. The plan was to clear a series of small hamlets in succession en route to the city itself. The first hamlet, Thon Lieu Coc Thuong, was cleared easily. Moving toward the second hamlet, Thon Que Chu, the battalion encountered a strong enemy force that was well entrenched. Air and artillery supported the battalion as it attacked and captured the hamlet against strong resistance. The battalion dug in for the night and prepared to resume the attack against the third hamlet, Thon La Chu, on 4 February.

Unknown to the 2-12th Cavalry, Thon La Chu was the headquarters of the Tri Thien-Hue Front, and it was defended by the NVA's 5th Infantry Regiment. Also important, the hamlet sat astride the NVA's primary supply route to the regiments fighting in Hue. The NVA was determined that 2-12th Cavalry's attack would not succeed; the NVA would destroy the threat to the Hue LOCs.

As the 2-12th Cavalry prepared to resume the attack on 4 February, the NVA fiercely counterattacked with all three battalions of the 5th NVA Regiment. It soon became apparent that 2-12th Cavalry could not continue to attack. As the day continued and the NVA pressure steadily increased, the battalion began to lose its position in Thon Que Chu. To avoid destruction, the battalion broke contact.

As darkness fell on 4 February, 2-12th Cavalry broke contact with the NVA. However, instead of retreating north back to its start line, the battalion moved west and then south to good defensive high ground deeper in the NVA AO. Eleven hours later the battalion was set in its new defensive position.

At dawn on 5 February, 2-12th Cavalry was established 5 kilometers west of Hue. The battalion soon observed enemy forces and supplies moving toward Hue. From its high ground position, the battalion directed artillery and air strikes against the NVA forces. By its bold move to bypass the 5th NVA Regiment, the battalion held perfect position to direct fires on the primary NVA supply line into Hue. These fires were the first step toward isolating the NVA in Hue.

The fires controlled by the 2-12th Cavalry shut down the NVA LOCs into Hue during the daytime. However, under the cover of darkness supplies and reinforcements were still entering the city (see Figure 6-15). The isolation of the NVA in Hue required the capture of Thong La Chu. The problem facing American forces was concentrating combat power against the NVA. All US units at this time were actively engaged against the numerous NVA attacks that constituted the NVA's 31 January Tet Offensive.

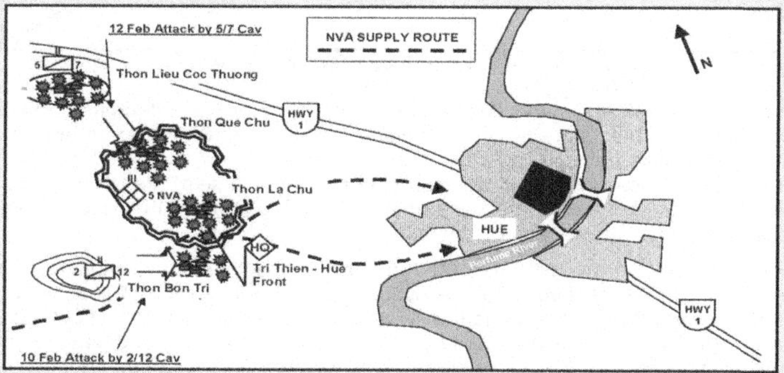

Figure 6-15. Subsequent Attack To Isolate Hue

The first additional American unit was not available until 12 February when the 5-7th Cavalry attacked Thong Que Chu much like the 2-12th Cavalry had attacked previously. The 5-7th Cavalry had even less success against the totally alert 5th NVA. The 5-7th Cavalry was forced to occupy defensive positions in Thon Lieu Coc Thuong and await the build up of combat power before it could continue to attack. In the interval, 2-12th Cavalry had moved off the high ground and captured the hamlet of Thon Bon Tri, south of the 5th NVA Regiment.

On 21 February, the 1st Cavalry Division had moved enough resources to the area to launch an effective attack to isolate Hue (see Figure 6-16 on page 6-24). In addition to the 5-7th and 2-12th Cavalry, the 1-7th Cavalry arrived in the AO and the 2-501st Airborne Infantry of the 101st Airborne Division was attached. On 21 February, after a combined artillery, air, and naval gunfire bombardment, the four battalions attacked the Thon La Chu stronghold. Elements of the 5th NVA Regiment were either destroyed in place or fled northeast. The next day resistance in Hue was noticeably lighter. US Marine and ARVN units began the last phase of fighting to recapture the Imperial Palace. On 26 February, the North Vietnamese flag was removed from the Citadel and the ARVN I Corps declared the city secured.

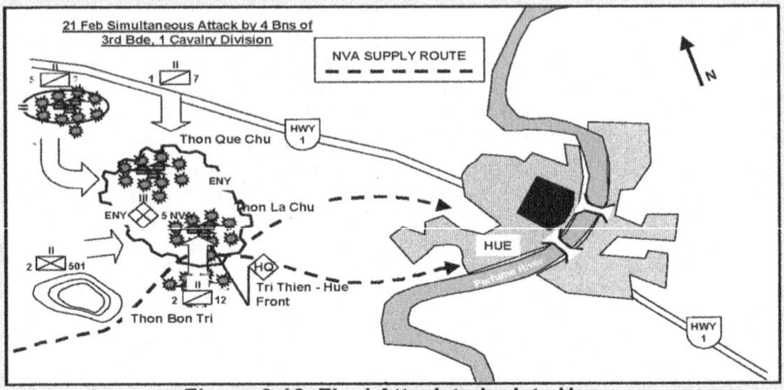

Figure 6-16. Final Attack to Isolate Hue

The actions of the 1st Cavalry Division forces northwest of Hue demonstrated the importance and the difficulty of isolating an enemy fighting in an urban area. Isolating Hue was difficult not only because of the dispersion and surprise with which the Tet Offensive caught US forces, but also because of the tenacity of the NVA. At least one-third of the combat power of the NVA in the Hue AO was focused on maintaining access to the city.

Hue's isolation had an immediate and important, if not decisive, impact on the operations. It not only resulted in restriction and then elimination of supplies and reinforcements, but it also immediately impacted the conduct of the defending NVA forces. Isolation caused an immediate drop in NVA morale and changed the nature of the defense. Once the enemy was isolated from external support and retreat, the objective of the NVA in the city changed from defending to avoiding destruction and attempting to infiltrate out of the city.

Direct Action by Special Operations Forces

6-61. Although SOF in urban offensive operations will likely conduct essential reconnaissance, they also have a direct action capability to shape the offensive operation (see Figure 6-17). Special Forces and Rangers can use direct action capabilities to attack targets to help isolate the urban area or to directly support decisive actions subsequently or simultaneously executed by conventional forces. Successful attacks against

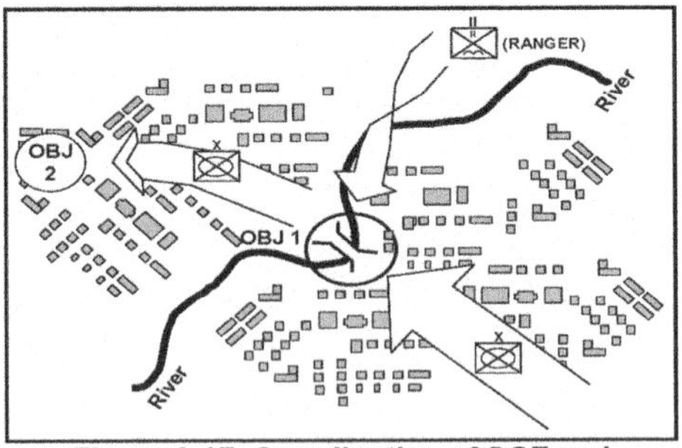

Figure 6-17. Coordination of SOF and Conventional Capabilities

urban infrastructure, such as transportation or communications centers, further the area's physical and electronic isolation. Direct action against command centers, logistics bases, and air defense assets can contribute to the success of conventional attacks by destroying or disrupting key threat capabilities. Direct action can also secure key targets such as airports, power stations, and television stations necessary for subsequent operations. Direct action by Special Forces and Rangers in these operations can reduce potential damage to the target or noncombatant casualties.

Information Operations

6-62. Regardless of how Army forces physically isolate the urban area, they combine physical isolation with IO to electronically isolate the threat and undermine his morale. Electronic isolation will cut off communications between forces in the urban area from their higher command to deny both from knowing the other's status. IO combined with isolation may persuade the threat's higher command or leadership that its forces located in the urban area are defeated. Thus, the command or leadership's intentions to break through to the besieged threat forces may be affected. PSYOP can undermine the morale of the threat in the area and reinforce electronic isolation and perceptions of abandonment. IO can be used to reduce any loyalty the civil population may have to the threat. IO can also ensure that civilians have the information that minimizes their exposure to combat and, as a result, overall noncombatant casualties. In addition, IO aim to deceive the threat regarding the time and place of Army force operations and intentions.

Detailed Leader Reconnaissance

6-63. Army commanders clearly see the urban environment to understand the challenges facing their brigades, battalions, companies, platoons, and squads. Urban terrain can be deceptive until viewed from the soldier's perspective. Commanders are responsible to intimately know the conditions to allocate resources effectively to subordinate units. Often, particularly at battalion level and above, commanders will not be able to command and control dispersed forces from positions forward, but be forced by the terrain to rely on semifixed command posts. Detailed leader reconnaissance of the AO by commanders, their staff, and their subordinates before the mission can compensate for this challenge. This reconnaissance will give commanders a personal feel for the challenges of the terrain and will facilitate more accurate planning and better decisionmaking during operations.

Mission Orders

6-64. Before contact, commanders mitigate some terrain challenges to effective C2 using mission orders. Subordinates have mission orders to take advantage of opportunities before C2 systems can adversely

> *Often what seems to be the correct decision at one level of command may be otherwise at other echelons. It is essential that leaders consider not only the perspective of their own unit, but that of other relevant participants as well, to include the enemy, adjacent friendly units, higher headquarters, and noncombatants.*
>
> Lesson Number 18
> *An Attack on Duffer's Downtown*

impact the environment. To see the battle and provide effective and timely direction, tactical leaders will follow closely behind units as they assault buildings, floors, and rooms. Thus, only the most mobile INFOSYS can accompany tactical leaders into combat, and they will suffer the degrading effects of the environment. Mission orders permit rapid and decisive execution without commanders intervening at battalion level and above. Higher-level commanders facilitate mission orders through their subordinates by articulating their desired end state, clearly stating their intent, and building flexibility into the overall plan.

Effective Task Organization

6-65. Commanders can shape urban offensive operations through effective and innovative task organization. Combined arms, starting with an infantry base, are essential to success and may be an asymmetric means of defeating an urban threat. Urban attacks will quickly break down into noncontiguous firefights between small units. To achieve the tactical agility for mission success in this nonlinear environment, many Army capabilities are task organized down to the company, platoon, and squad levels. Infantry provides the decisive capability to enter buildings and other structures to ensure threat destruction. Tanks, gun systems, and fighting vehicles provide additional mobility, direct firepower, and protection. Field artillery provides the indirect (and if necessary, direct) firepower. Such mobility and firepower create the conditions necessary for the dismounted infantry to close with and destroy a covered threat in an urban defense. When a threat skillfully uses the urban area to limit ground maneuver, vertical envelopment or aerial attack using precision-guided munitions from Army aviation may circumvent his defenses and achieve necessary effects. Generally, ground systems used *within* the urban area will not be able to operate independently from dismounted infantry. The infantry will be required to protect armor and mechanized systems from close antiarmor weapons, particularly when those weapons are used from rooftops and basements.

6-66. In urban offensive operations, direct fire support can be critical. Armor vehicle munitions types do not always achieve decisive effects against some urban structures. In some cases, field artillery high explosive munitions work better than armor for direct fire support of infantry. Large caliber (105 or 155mm) high explosives directly fired at a structure often produce a more severe shock effect than tank and fighting vehicle cannon and machine guns produce. This is not just a weapon but a weapon system. As such, artillery may be placed under tactical control (TACON) of maneuver commanders, such as a platoon of three guns TACON to a company or a battery to a battalion, not just one gun to a company or other maneuver unit. Self-propelled artillery has some of the mobility characteristics of armor; however, it provides minimal ballistic protection from fragmentation for the crew. Although these systems seem formidable, they provide less crew protection than a Bradley fighting vehicle, for example, and contain large amounts of onboard ammunition and propellant. They are susceptible to catastrophic destruction by heavy automatic weapons, light cannon, and antitank fire. Therefore, infantry units carefully secure and protect these systems (even more so than armored vehicles) when employed in urban offensive operations, particularly when forward in the direct fire role.

6-67. Army aviation may also be inventively task organized. It can support urban operations with lift, attack, and reconnaissance capabilities. Tactical commanders down to company may use all these capabilities to positively influence ground close combat. Army attack and reconnaissance aircraft can provide flank security for attacking ground forces. Attack aircraft may also provide direct fire support to individual platoons or squads. Lift may move entire battalions as part of brigade operations, or it may move single squads to a position of advantage (such as a roof) as part of a small unit assault. Army aviation can assist with C2 by providing airborne retransmission capability, airborne command posts, and the confirmed status and position of friendly forces. However, Army aviation is a limited and high-value asset; commanders review its use in innovative task organizations. It is particularly vulnerable to urban air defense threats unless used over terrain secured by ground forces. From these positions, aircraft can use enhanced sensors to conduct reconnaissance and use precision weapons with standoff capability.

Creative Task Organization: Using Artillery in the Direct Fire Role

Task organizing artillery to permit its use in a direct fire role demonstrates the innovative task organization required for urban operations. The following provides three historical examples of task organizing and using field artillery for a direct fire role.

In 1944, US Army units of the 1st Infantry Division were assigned to attack and seize the German city of Aachen. The city's internal defense included bunkers designed to serve as air raid shelters. These positions, buildings of stone, were impervious to direct fire tank weapons, demolitions, and small arms. To reduce the positions, the 1st Infantry Division relied on the artillery's direct fire.

Field artillery used this way had physical and psychological effects on the defenders. The 26th Regimental Combat Team's history of the battle describes the German reaction to the artillery pieces:

> *The chief shock to the defenders, Colonel Wilck (Aachen defense commander) said, came from the self-propelled 155s and tanks. The colonel spoke with considerable consternation of the 155mm self-propelled rifles. A shell from one of the guns, he said, pierced three houses completely before exploding and wrecking a fourth.*

The 26th Infantry Regiment also described how the artillery, one piece attached to each assaulting infantry battalion, helped the infantry to penetrate buildings.

> *With solid blocks of buildings comprising most of the city, there wasn't any easy way to get at the Germans in the buildings. The eight-inch gun solved the problem. Beginning on the eastern outskirts the gun would plow a round into the side of the built up block of buildings at about ground level. One shell would usually open an entrance into the first tier of floors, i.e. the first building. Then several more shells were fired through the first hole. Thus a tunnel would be rapidly made all the way to the next cross street. Soldiers could then rush the newly formed entrance, clear the upper floors with hand grenades and rifles and then move on to the next building to repeat the process. When a block or*

square, was thus completely cleared of Germans—soldiers, skulkers, or even snipers—the next square was treated in the same way, working forward square by square, right and left, thereby avoiding nearly all exposure in the streets.

In 1982, Israeli forces invaded southern Lebanon to destroy base camps of the Palestine Liberation Organization (PLO). This operation involved significant fighting in urban areas including major operations in Beirut. Artillery, firing in a direct fire role, played a major part of the tactical solution. Artillery was particularly effective in the 33-day siege of Beirut. During this siege, Israeli forces used artillery in its traditional role as well as in the direct fire role.

The Israeli army was committed to a policy of disproportionate response during the Beirut siege. When fired on with small arms, crew-served weapons, tanks, or indirect artillery, the Israeli forces responded with intense, high-caliber direct and indirect fire from tanks and artillery positioned around the city. Many firing positions were on heights to the south and southwest that dominated much of the city. These positions had almost unrestricted fields of view. Israeli artillery fired from these positions directly into high-rise buildings concealing PLO gunners and snipers. The artillery, using direct fire, destroyed entire floors, collapsed floors on top of each other, and completely removed some upper floors. Such a response, as in Aachen in 1944, had as much a psychological impact as it did a physical impact on the PLO defenders.

In the early hours of 20 December 1989, the US launched OPERATION JUST CAUSE. One of this operation's objectives was removing the Panamanian dictator, Manuel Noriega. US forces carefully planned using all fires before the operation to minimize casualties and collateral damage. Part of this detailed fire planning called for applying artillery in a direct fire role.

The Panamanian Defense Force's (PDF) 5th Rifle Company based at Fort Amador was one of the key objectives of US forces at the start of hostilities. This unit was high priority because it was the closest PDF unit to Noriega's headquarters, the Comandancia. US forces expected the 500-man company to react first to OPERATION JUST CAUSE by reinforcing the defense of the Comandancia. It also posed a threat to US military dependents housed at Fort Amador.

To quickly neutralize this force, the US assembled a three-company force composed of A, B, and headquarters elements of 1-508th Infantry (Airborne), supported by 105mm towed howitzers of 320th Field Artillery and M113 armored personnel carriers. The howitzers and the personnel carriers were covertly prepositioned at the fort. At approximately 0100, helicopters transported the two airborne rifle companies into position. The howitzers then suppressed any personnel in the PDF-controlled buildings on Fort Amador while demonstrating the firepower of the US task force. They used direct fire into the PDF barracks. The impact of the 105mm high explosives and .50-caliber fire from the M113s convinced the PDF infantry to give up after token resistance. Following the direct fire, US infantry assaulted and cleared the dozen PDF buildings, finding that most occupants had fled or surrendered. For more details of OPERATION JUST CAUSE, see Applying the Urban Operational Framework: Panama in Chapter 5.

The three examples cited indicate the importance of the innovative task organization of artillery and its use in the direct fire role. Using artillery helps overcome some challenges of offensive operations in the urban environment, and it has an important psychological effect on a defending threat. Such task organization takes a traditional tool of a higher-level tactical commander and uses it to directly influence the company-level battle. This philosophy of task organization can be applied to other types of forces—not just artillery. PSYOP teams, interpreters, CA specialists, armor, and reconnaissance teams may require task organization different from traditional organization. The compartmented urban environment drives the requirements for these assets lower in the tactical scheme than in open operations. Consequently, commanders understand and account for more of these assets for UO than for operations in less restrictive environments.

Shaping Attacks

6-68. In a large urban area, the defending threat cannot be strong everywhere. Shaping operations can also take the form of attacks against vulnerable positions to force the threat to maneuver and redeploy his forces in the urban area. This prevents him from merely defending from prepared positions. Forcing the threat to move negates many of the defensive advantages of urban terrain, confirms his dispositions, exposes vulnerable flanks, and permits target acquisition and engagement with precision standoff fires.

6-69. A critical shaping operation in urban offensive operations is usually an initial attack to seize a foothold. Once Army forces establish this foothold, they accrue some of the defensive advantages of urban terrain. From this protected location, Army forces continue offensive operations and have a position of advantage against neighboring threat defensive positions.

DOMINATE

6-70. Commanders may employ several methods to dominate the urban area during offensive operations. These include—

while the area remains relatively undefended—essentially preempting an effective defense. This opportunity occurs when the urban area is well to the rear of defending threat forces or before the onset of hostilities. Under such conditions, an attack requires striking deep behind threat forces or striking quickly with little time for the threat to make deliberate preparations. Attacks under such conditions may entail significant risk; the potential benefit of audacious offensive operations may be well worth possible losses. Such attacks can be accomplished three ways (and their combinations):

counterattack north and recapture Seoul. The disadvantage of this course of action was that the counterattack would be through hundreds of miles of mountainous terrain, through several major cities, and across numerous rivers. The North Korean People's Army (NKPA) would be able to fall back from one mountain defense and one river line to another and would have weeks or months to prepare the defenses of Seoul.

MacArthur's solution was to delay a conventional counterattack, boldly maneuver by sea deep into the flanks of the NKPA, land on the Korean west coast at Inchon, and quickly seize Seoul before the enemy could react (see Figure 6-18). This course of action had numerous disadvantages: achieving surprise; assembling a landing force trained in amphibious operations; few good landing sites; supplying a large force once landed; and needing to simultaneously continue to defend and prepare a counterattack with the Eighth Army from Pusan. MacArthur understood the challenges of the operation but believed that the potential value compensated for the risk.

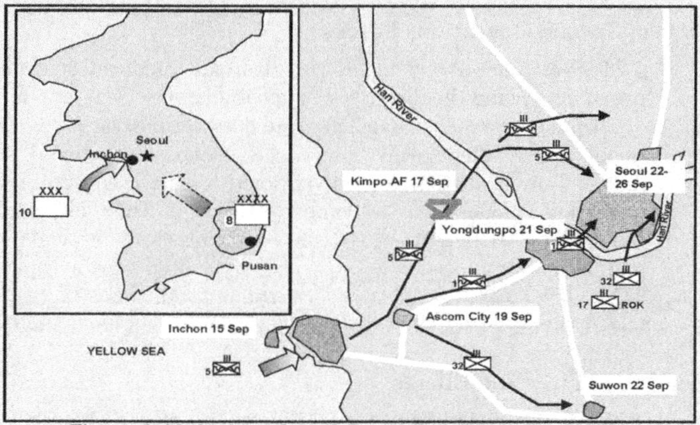

Figure 6-18. Inchon-Seoul Campaign, September 1950

The potential value of a quick assault on Seoul through a deep amphibious operation was immense. The operation would trap the bulk of the NKPA in South Korea and facilitate its destruction. It would also capture Seoul before the NKPA could prepare a defense. The quick recapture would immeasurably impact the morale of UN forces and South Korea. MacArthur felt the payoff of success justified the risks and challenges of the operation.

On 15 September, elements of the 1st Marine Division landed in Inchon taking the NKPA completely by surprise. On D+1 they were joined by elements of the 7th Infantry Division, both divisions under the control of US X Corps. Naval gunfire and Marine close air support flying from Navy aircraft carriers supported the landings. Seven days later, the 5th Marine Regiment had battled 25 miles to the outskirts of Seoul and began attacking the city from the north. Three days later, on 25 September, the 1st Marine Division and elements of the 7th Infantry Division had defeated the 10,000 defenders of the NKPA's 18th Rifle Division.

> Most of MacArthur's predictions for the Inchon-Seoul campaign were accurate. Superb staff work, excellent logistics operations, and unmatched training permitted the UN forces to assemble a trained landing force, land it over one of the most difficult shores in the history of military amphibious operations, and keep it supported. More importantly, the operation achieved total strategic and operational surprise. Seoul was recaptured quickly and, although resistance was fierce, the NKPA could not react fast enough to influence the outcome. In conjunction with the Eighth Army counterattack from Pusan, 75 percent of the NKPA was destroyed. The Inchon-Seoul campaign was an important turning point in the war, and had not the Chinese People's Army intervened two months later, it would have been the decisive campaign of the war.

Appropriate Use of Special Operations Forces

6-74. Sometimes Army forces can dominate (instead of simply shape) the urban area using the direct action capability of SOF. When the threat fails to develop a comprehensive defense and does not possess large, capable conventional forces, then Army forces can achieve operational surprise. Commanders, by synchronizing conventional and SOF effects, may actively control offensive operations to dominate the area. Then, although SOF may be the primary striking force, conventional forces still are available to reinforce and assume the mission because of SOF limited logistic capability. OPERATION JUST CAUSE offers several examples of this type of synchronization (see Applying the Urban Operational Framework: Panama in Chapter 5).

Precise Application of Fires and Effects

6-75. Precisely applied fires and the massed effects of combat power characterize successful urban attacks. The fires can be direct fire from combined heavy or light ground teams; direct or indirect fires from supporting Army aviation standing off from the target and any possible air defense threat; or precision indirect fires from conventional tube artillery. All efforts strive to reduce collateral damage around the point of attack, consistent with mission success. Forces use fires to deny the threat the ability to maneuver in the urban area and to destroy the threat when he attempts to maneuver. When the threat exposes himself by moving, the environment no longer protects him, and fires can effectively engage him.

Proper Balance of Speed and Security

6-76. Attacking units balance speed and security. Forces secure flanks as units advance, control dominating terrain (buildings), evacuate civilians, and keep the integrity and synchronization of the combined arms team. Obstacles are anticipated and rapidly breached. Commanders choose avenues of approach to—

6-77. Army aviation is one resource to protect flanks. Another resource is engineers who seal off surface and subsurface entries and avenues along the flanks of the attack. Finally, as in all offensive operations, ground and air cavalry are ideal mobile forces to perform security in an economy of force role along flanks allowing decisive forces more freedom of maneuver.

TRANSITION

6-78. Effective transitioning allows commanders to continue the full range of Army operations in the urban area and elsewhere in the AO without unnecessary delays. Commanders accomplish this by thorough planning—including appropriate branches and sequels (revised as the situation changes)—that gives adequate consideration to postoffensive organizational, training, psychological, and civilian requirements. If properly prepared, commanders can anticipate potential mission changes and defeat a counterattack.

Early and Concurrent Transition Planning

6-79. Commanders can ensure smooth transitions of urban offensive operations by planning for postoffensive operations early. Based on the mission envisioned, they determine which subordinates and what type of force structure to use. Postoffensive missions, like all urban operations, encompass the full range of Army operations. At the successful conclusion of offensive operations, Army forces transition to some type of stability operation conducted concurrently with support operations. Commanders may leave the subordinate unit in place to execute the new mission, may reorganize the subordinate unit for the mission, or relieve the unit that just completed offensive operations with a new unit.

6-80. **Changes to Task Organization.** Commanders consider the organization of forces following offensive operations. Hostile civilians may require significant combat forces or military police forces to maintain stability. Friendly civilians may require a minimum of military police or combat forces, but significant logistic support. Commanders carefully consider the urban situation before deciding how to use combat forces that recently participated in a high-intensity offensive operation.

6-81. **Training and Psychological Considerations.** Many Army combat tasks may not support follow-on stability operations or support operations without considerable modification (if at all). Often, noncombat skills—not normally part of a unit's mission essential task list such as negotiating or mediating skills—will be required. However, the greatest modification required applies to each soldier's mental outlook. Forces that transition directly from combat to stability operations may not be psychologically prepared for a rapid and drastic change of mission. Commanders cannot expect troops who have just completed high-intensity offensive operations to exercise the sensitivity and judgment required in most stability operations. This especially applies if the population is hostile to Army forces. If combat forces assist in stability operations, particularly in hostile civilian situations, they should not have recent experience in high-intensity urban operations and they should have trained for the mission.

6-82. **Return to Civil Agencies.** Commanders also have the role of transitioning aspects of the urban offensive operation to civil, allied, NGOs, and other agencies as appropriate. Planning for transition is detailed and aims to quickly return as much civil control of the area as is feasible after the attack. Beyond civil control, civil agencies and NGOs assume tasks as completely and as rapidly as possible. These organizations are consulted and integrated into the planning process as early as possible. Commanders begin planning for transition when planning for offensive operations. They consider relinquishing control of urban areas to civil government, law enforcement, or NGOs before completing offensive operations. During the conduct of urban operations, these transition operations are closely synchronized with the execution of the attack.

Preparation for Potential Mission Changes

6-83. In urban offensive operations, like other offensive missions, the change in mission after a successful urban attack may be to a hasty defense or a continuation of offensive operations outside the area. However, in urban offensive operations the mission will just as likely change to a support or stability mission. This is particularly true if the unit has had special training and is task organized for urban operations. Stryker brigades, already optimized for urban environment by organization and training, are even more likely to remain after completing the mission.

6-84. Even more challenging than transition at the end of the mission is transition during the conduct of the mission. Soldiers may have a difficultly transitioning from stability to support, to offense and defense, and back again multiple times during an urban offensive operation. Soldiers may apply the tactics, techniques, and procedures of urban offensive operations to the stability or support missions with potentially disastrous results. Commanders need to segregate missions in time and space, and, if sufficient forces exist, by unit. Units below battalion level cannot have rapid mission changes from combat to noncombat or be expected to conduct more than one type of urban mission simultaneously. Whenever feasible given mission requirements, commanders permanently designate units to conduct stability or support tasks and not rotate units between offensive and defensive tasks to stability or support tasks.

Preparation to Defeat Counterattacks

6-85. Immediately after the conduct of successful urban operations, units are alert to rapid and violent counterattack. Rapid and decisive counterattacking in urban operations is extremely important to the defense because a quick counterattack can regain terrain before the offensive forces have consolidated and assumed the advantages of defending on urban terrain. Delaying a counterattack in UO, even for a few minutes, permits the environment's advantages to shift to the successful attacker. Thus, attacking units anticipate this reaction during planning and are prepared to defeat it during execution.

Chapter 7

Urban Defensive Operations

Generally, a modern city magnifies the power of the defender and robs the attacker or his advantages in firepower and mobility. A city can ingest an invading army, paralyze it for weeks on end, and grind it down to a state of ineffectiveness.

"Military Operations on Urbanized Terrain:
The 2d Battalion, 26th Infantry, at Aachen, October 1944"

The skillful defense of an urban area can decisively affect a campaign. The urban area offers many advantages to defending forces. An adroit defender can use the advantages of the urban environment to negate combat power disparities, blunt the tempo of an attack, attrit threat forces, and sap the morale of attacking troops. The defender gains an opportunity to concentrate resources, reconstitute attrited units, and transition to the offense. A successful defense of an urban area can also deny the threat vital resources. Defense in the urban environment is an essential Army capability and can significantly affect the outcome of entire campaigns and the achievement of national objectives.

PURPOSE OF URBAN DEFENSIVE OPERATIONS

7-1. Army forces defend urban areas for various reasons: defeating a threat attack, buying time, economizing forces, protecting an ally's political institutions and economic infrastructure, protecting an urban population, shaping conditions for decisive offensive operations, and shaping conditions for executing stability operations or support operations. During force projection operations, urban areas may be used as initial lodgment areas that Army commanders may need to defend at the outset until they build sufficient combat power. Usually two or more of these purposes apply to the urban defense. Urban defensive operations provide commanders great opportunities

CONTENTS

to turn the environment's characteristics to the advantage of Army forces. Urban areas are ideal for defensive operations and greatly enhance the combat power of defending units.

CHARACTERISTICS OF URBAN DEFENSIVE OPERATIONS

7-2. There are five general characteristics of the successful defense: preparation, security, disruption, massing effects, and flexibility. All apply to the successful urban defense and to the higher commander supporting a subordinate defending in the urban area.

PREPARATION

7-3. The urban area suits the defense since the area's physical characteristics naturally enhance the combat power of defending units. These characteristics include protection, obstacles, and concealment. Urban terrain provides superb defensive positions with minimum preparation. With deliberate preparation, urban defensive positions can rapidly become strong points.

7-4. One primary characteristic of urban terrain that enhances the defense is protection. With little or no advance preparation, buildings, subsurface structures, and walls protect soldiers from direct and indirect fire, interdict indirect fire, limit observation, and limit engagement ranges. Nearly all buildings provide some ballistic protection from direct and indirect fire. Mason and stone buildings with basements and cellars can protect soldiers from most fires except the largest caliber or tonnage bomb. Minimal additional preparation turns them into formidable, defensive strong points.

7-5. Buildings in urban areas, because of their height and close proximity, also can protect soldiers by masking them from indirect fire. The height of a building may interdict the flight path of an artillery round, rocket, missile, or bomb at a point short of the intended target. Masking protects static defending forces and protects forces moving along routes bordered with tall buildings that form urban "canyons". These protected routes can be used for logistics, counterattacks, and maneuver.

7-6. Structurally significant buildings in an urban area can create major obstacles to maneuver. These obstacles immediately canalize maneuver into existing streets and routes without any preparation by the defense. These obstacles then become kill zones for well-positioned and sited defensive forces. Minimal obstacle construction as point obstacles blocking streets and routes can further restrict the maneuver options of the attacking force. Rubble from structures collapsing into streets after fires (intentional or unintentional) can also block routes.

7-7. Buildings also conceal the location, disposition, and intent of the defense. They limit visual observation to the external perimeter of the urban area. They degrade radar and electronic position identifiers and decrease the utility of overhead imagery. The physical aspect of the urban environment greatly enhances the defense by degrading the opposition's intelligence, surveillance, and reconnaissance (ISR) capabilities. Buildings can conceal static defensive positions and the maneuver of defensive forces in the urban area. Although the environment constrains defensive mobility in much the same manner as offensive mobility, the defense has the time and opportunity to conduct careful reconnaissance and select and prepare routes. This gives the defender the ability to move reserves, maneuver counterattack forces, and plan logistics without observation. Careful preparation provides the defense a mobility advantage over attacking forces.

SECURITY

7-8. The urban area can be an advantage or a disadvantage to the security of defending forces. This largely depends on the nature of the human dimension of the environment. If the population is evacuated or allied with Army forces, then the environment may assist in the security of defending Army forces. However, if the population is present and hostile, then the environment may make security difficult.

7-9. The physical aspects of the urban environment, uninfluenced by the human dimension, may assist in the security of defending Army forces. The combat power of small security forces manning observation posts is greatly enhanced. Forces can more easily restrict and monitor avenues of approach for threat reconnaissance. Defending forces positioned mostly in structures are difficult to locate.

7-10. The physical aspects of the environment may also present some security challenges, primarily with observation. The compartmented terrain limits the field of observation from any one point. The defense may require more security forces to adequately observe the mounted and dismounted avenues to prevent infiltration. Threat forces that successfully infiltrate will be more difficult to locate. These forces will gain numerous hide positions for small reconnaissance units in complex terrain and the effect the terrain has in masking electronic signatures.

7-11. Friendly civilians in the urban area can help identify threat forces attempting to conduct reconnaissance. Civilian activity will also help to mask defense preparations. However, a hostile element of the population may pass intelligence information to the threat. They may assist threat reconnaissance to infiltrate the urban area or provide guides, manpower, or resource support for threat forces. Commanders take measures to ensure strict control of hostile populations. If resources permit, commanders may consider removing potentially hostile civilians from the area.

DISRUPTION

7-12. The urban environment's attributes assist defending Army forces to disrupt the attacker. It does this through compartmentalization, inhibiting command and control (C2), and facilitating counterattacks.

7-13. The physical aspects of the urban area force the attacking threat into compartmented urban canyons that make mutual support between attacking threat columns difficult. Shifting resources from one portion of the threat attack to another is also difficult. Physically, the urban area disrupts tactical communications making synchronization of combat power difficult.

7-14. The urban terrain hinders the mobility capabilities of the defense. However, careful planning, preparation, and rehearsals can facilitate more rapid movement of larger forces. Defending forces can assemble counterattacks undetected, move them along covered and concealed routes, and achieve surprise at the point of the counterattack. Attacking forces, using the compartmented terrain, often leave forward elements in position to be isolated or expose long and vulnerable flanks to friendly counterattack and interdiction.

MASSING EFFECTS

7-15. The urban environment allows defenders to better protect their centers of gravity and decisive points. The restrictive terrain reduces the attacker's maneuver options. Defenders can position forces in protected and mutually supportive positions oriented on deadly engagement areas. Relatively few well-positioned defenders can generate significant combat power. Without the positional advantage and the corresponding protective effects of the terrain, attacking forces often mass numbers to achieve the necessary combat power.

7-16. Knowing the complex terrain permits defending forces to plan engagement areas that maximize the effects of their combat power. Defending forces can remove fences, walls, rooftops, and even entire buildings to facilitate fields of fire and unmask indirect fire flight paths. Forces carefully choose firing positions for indirect fire systems so that flight paths travel between buildings into engagement areas. By leveraging this knowledge of the terrain, numerically inferior defenders can synchronize devastating fires on offensive forces that are forced by terrain and reinforcing obstacles to mass in confined spaces where fires can have the greatest effect.

FLEXIBILITY

7-17. Defensive flexibility results from detailed planning and mental agility. Defensive planning flexibility forms branches and sequels that include alternate and subsequent positions and emphasize counterattack options. The urban area facilitates defensive flexibility because the urban terrain can be quickly adapted for defensive operations with little or no preparation. The effect is similar to having multiple, prepared positions on nearly every possible approach. The urban area can also permit rapid, covered movement on interior lines. This permits swift movement to and occupation of strong defensive positions with little or no preparation. The defense also has more flexibility since defenders often know and better understand the urban terrain's effects on operations. Normally, defenders will not get lost as easily, will know complex lines of sight and masking effects, and will best understand the ballistic characteristics of individual structures.

7-18. Mental agility allows commanders to see that the best urban defense may actually be to defend *outside* of the area. Such a defense mitigates the danger to the urban population and potentially reduces collateral damage. It takes advantage of Army long-range engagement capabilities and denies the threat the opportunity to "hug" Army forces or noncombatants as protection from fires. This defense may be appropriate when Army forces have enough resources to defend more open terrain, when time permits deploying extensive obstacles and constructing protected positions, and when natural terrain such as river obstacles aids the defense.

URBAN DEFENSIVE OPERATIONS AND BATTLEFIELD ORGANIZATION

7-19. Urban defensive operations are organized within the overall doctrinal framework of sustaining, shaping, and decisive operations. The success of urban defense depends on each operation, but commanders synchronize these simultaneous operations as one action. Sustaining operations in defensive urban operations (UO) ensure freedom of action. Critically, urban sustaining operations ensure security of the lines of communications and establish effective movement control. Shaping operations in defensive UO create the conditions for decisive operations. Shaping operations vary greatly depending on the type of defense. For example, in a mobile defense the shaping operation may be the fixing force. In contrast, in an area defense the fixed defense may be the decisive operation. In the urban defense, decisive operations focus on accomplishing the commander's mission. The decisive operation may not defeat the threat's main effort, and it may not prevent threat occupation of large portions of the urban area if those tasks are not essential to mission accomplishment. For example, if the defense's objective is to protect a critical communications node, then, depending on the commander's overall intent, threat actions to secure an airfield elsewhere may not be important.

TYPES OF URBAN DEFENSE

7-20. Commanders view urban area defensive operations two ways: as conducting a major defensive operation with an urban area in their area of operations and as defending entirely in an urban area.

AREA DEFENSE

7-21. At the operational level, an area defense may include both urban areas and open maneuver areas. The most common defense in an urban area and the most suitable for the characteristics of this distinct environment is the area defense. As a defensive operation, the area defense concentrates on denying threat forces access to designated terrain for a specific time rather than destroying the threat outright. Although an area defense in an urban area does not directly seek to destroy or defeat attacking threat forces, as an objective it does aim to force culmination of the threat's attack. The urban area defense often works effectively to exhaust threat resources and shape conditions for a transition to offensive operations. The urban area may also be used as a strong point to force threat movement in a different direction or to fix threat forces as part of a large, mobile defense taking place in the area of operations (AO) outside the urban area (see paragraphs 7-22 to 7-24).

MOBILE DEFENSE

7-22. A mobile defense can operate *in* an urban area but only under specific conditions. It focuses on destroying or defeating the threat through a decisive attack by a striking force. It requires the defender to have greater mobility than the attacker. To shape a mobility advantage, the urban defender effectively uses the terrain and correctly task organizes his forces mobility. The principles of applying the mobile defense in the urban area remain the same: a small fixing force stops the threat and limits any ability to maneuver while a striking force quickly maneuvers and counterattacks to destroy him.

7-23. One key to executing a mobile defense in the urban area is to entice a threat force into the depths of the urban area where it begins to lose mobility options. A well-placed fixing force augmented with man-made obstacles and taking advantage of the naturally constrictive terrain can stop a much larger force. If the attacking force is largely mounted and armored, its mobility in the urban area may be reduced to less than that of dismounted infantry. In addition, if the attacking force's movement into the urban area is mounted and rapid, the commander's situational understanding also diminishes. Then the striking force, consisting of dismounted infantry forces, can execute the counterattack with surprise from multiple directions and dimensions (subsurface, surface, supersurface to include intrasurface, and airspace). Man-portable antiarmor weapons—firing from flanks and top down and supported by precision indirect fires from both organic and joint systems—can rapidly destroy the threat.

7-24. From the perspective of commanders of the major operation, the urban environment can help defending forces achieve a mobility advantage over an attacker in a broader sense. Defending commanders attempt to shape the battlefield so that the attacker commits significant resources into an urban area, where his maneuver capabilities are reduced (see Figure 7-1). A disproportionately small defending force, which relies on the defensive combat power advantages of the urban environment, reduces and fixes the attacker's maneuver capabilities. Other defending forces mass *outside* the urban area then strike the threat with a combined mobility and firepower advantage.

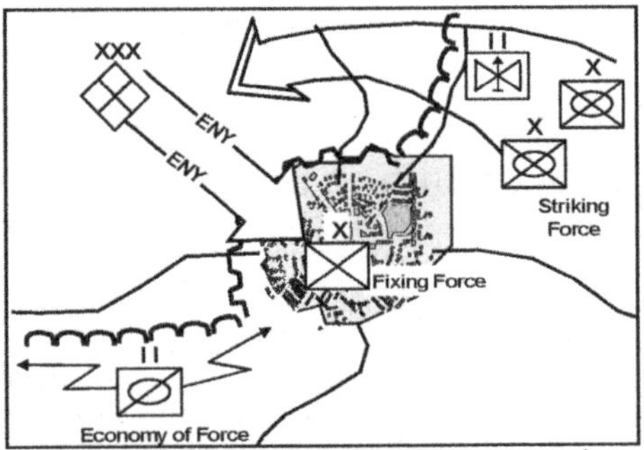

Figure 7-1. An Urban Area Incorporated Into a Larger Mobile Defense

Urban Defense in a Major Operation
Stalingrad – August 1942 to January 1943

The German and Soviet battle for Stalingrad in late 1942 and into early 1943 illustrates how a tactical urban area defense integrates into a larger mobile defense. Stalingrad was a battle fought on a huge scale: it involved army groups on both sides and thousands of square kilometers. Though the city was small, it remained the focus of both German and Soviet offensive and defensive operations during the six-month battle.

In the summer of 1942, the Germans launched a strategic offensive in southern Russia. Its goal was the valuable oil fields of the Caucasus. German forces turning south into the Caucasus exposed a vulnerable flank to Soviet forces positioned between the Don and Volga Rivers. For the German Caucasus operation to succeed, it had to destroy Soviet forces between the Don and Volga, establish a good defensive line, and capture Stalingrad. This city would anchor the German defense and interdict the critical flow of supplies from the Caspian Sea via the Volga River into central Russia. Stalingrad, by virtue of its name, also had important political and cultural value to the Germans and Soviets.

The opening phases of the German offensive were successful; German forces—the 6th Army and 4th Panzer Army—entered the outskirts of Stalingrad in late August 1942 (see Figure 7-2). After a month of intense fighting, the Germans possessed nearly 90 percent of the city. At this point, the 6th Army commanded all German forces in and around Stalingrad. The Soviet 62nd Army's defense was reduced to a front only a few hundred meters deep and a couple of kilometers long on the banks of the Volga. The Soviet defenses hinged on fortress-like concrete industrial buildings and the fanatical bravery and tenacity of soldiers and civilians fighting in the city's remains (see Figure 7-3).

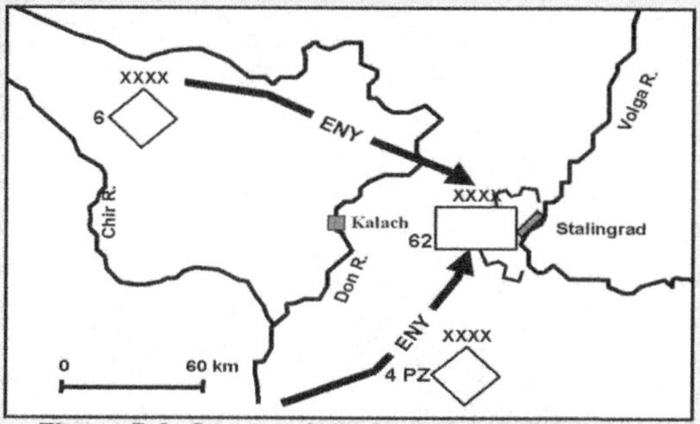

Figure 7-2. German Attacks to Seize Stalingrad

Beginning in mid-September, the Soviet command began looking at how to convert the defense of Stalingrad into an operational opportunity. During October and November, the 62nd Army held on to its toehold in Stalingrad. While maintaining the defense of the 62nd Army, the Soviets secretly began to build up strength on both flanks of the German 6th Army. The Germans increased their vulnerability by securing the German 6th Army's flanks with less capable Romanian, Hungarian, and Italian armies. Also, the 6th

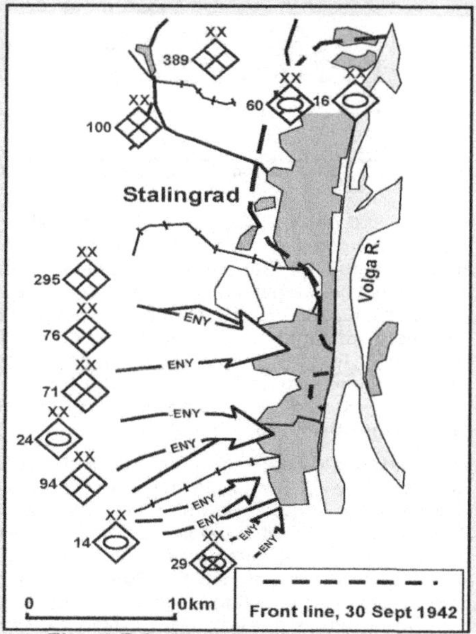

Figure 7-3. German Attacks to Seize Stalingrad, September 1942

Army moved powerful German divisions into the city and rotated with German divisions that were exhausted by urban combat.

On 19 November, the Soviets launched OPERATION URANUS that attacked two Romanian armies with seven Soviet armies. Simultaneously, the 8th Russian Army attacked to aid the 62nd Army in further fixing the German 6th Army. Within five days, the Soviet armies of the Don Front, Southwest Front, and Stalingrad Front met near the city of Kalach and sealed the fate of the German 6th Army's 300,000 troops in Stalingrad (see Figure 7-4).

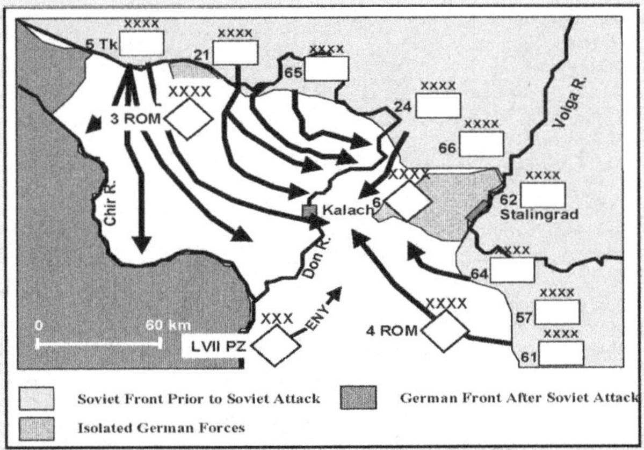

Figure 7-4. Soviet Attacks Trap German 6th Army

On the third day of the Soviet offensive, when encirclement seemed inevitable but not yet complete, the 6th Army commander asked permission to withdraw from the trap. The German high command denied permission believing that the Army could be supplied by air and then a renewed offensive could break through to the city. On 12 December, the German LVII Panzer Corps launched an offensive north to break through to Stalingrad. This offensive made progress until another Soviet offensive on 16 December forced its cancellation. This ended any hope of recovering Stalingrad and the 6th Army. On 31 January 1943, the 6th Army surrendered after sustaining losses of almost two-thirds of its strength. The Soviets took over 100,000 prisoners.

Many lessons emerge from the successful defense of Stalingrad. Tactically, the defense showed how using the terrain of a modern industrial city wisely could increase the combat power of an inferior, defending force and reduce the maneuver options of a mobile, modern attacking force. Another element in the Soviet's tactical success was the Germans' inability to isolate the defenders. The Germans never threatened the Soviet supply bases east of the Volga and, despite German air superiority, the Soviets continuously supplied and reinforced the 62nd Army across the Volga River. Also, Soviet artillery west of the river was able to fire in support of Soviet forces and was never threatened with ground attack.

At the operational level, the Soviets demonstrated a keen understanding of using an urban area within the context of a mobile defense. The 62nd Army's stubborn area defense of Stalingrad drew the bulk of the German combat power into the urban area where they were fixed by a smaller and quantitatively inferior defending force. This allowed the Soviets to build combat power outside the urban area. The Soviets set the conditions for a mobile defense by positioning powerful Soviet armor forces in open terrain outside the urban area against quantitatively inferior German allied forces. In OPERATION URANUS, the mobile defense's strike force destroyed the enemy outside the urban area and trapped the greater part of the best enemy formations inside the urban area. The trapped units were then subjected to dwindling resources and extensive psychological operations, further isolated into pockets, and defeated in detail.

RETROGRADE

7-25. A retrograde involves organized movement away from the threat. Retrograde operations include withdrawals, delays, and retirements. These defensive operations often occur in an urban environment. The urban environment enhances the defending force's ability to conduct retrograde operations successfully (see Figure 7-5).

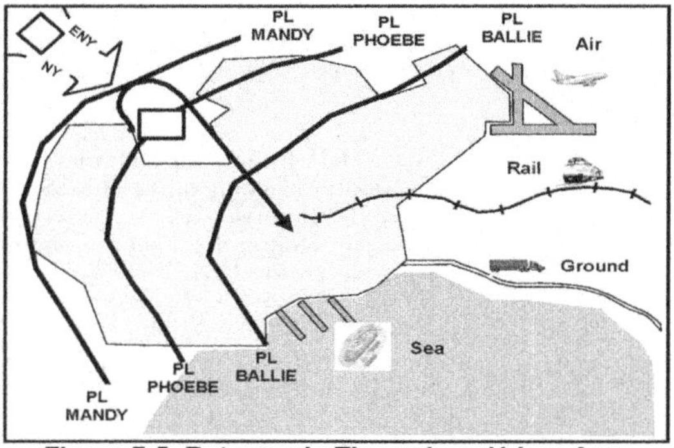

Figure 7-5. Retrograde Through an Urban Area

7-26. The cover and concealment afforded by the urban environment facilitates withdrawals where friendly forces attempt to break contact with the threat and move away. The environment also restricts threat reconnaissance, which is less able to detect friendly forces moving out of position, and presents excellent opportunities for deception actions. Finally, a small security force's ability to remain concealed until contact in the urban environment significantly slows threat attempts to regain contact once Army forces have broken contact and begun to move.

7-27. The urban environment's natural cover and concealment, as well as the compartmented effects, facilitates delays. Delays can effectively draw the threat into the urban area for subsequent counterattack or as an integral part of a withdrawal under threat pressure. Delaying units can quickly displace from one covered and concealed position to another; the repositioning options are vast. Compartmented effects force the attacking threat to move on well-defined and easily interdicted routes and limit the threat's ability to flank or bypass delaying positions.

7-28. The urban area's transportation and distribution network facilitates retiring forces that are not in contact. Properly used, the urban transportation system can quickly move large forces and associated resources, using port facilities, airfields, railheads, and well-developed road networks.

URBAN DEFENSIVE CONSIDERATIONS

> *What is the position about London? I have a very clear view that we should fight every inch of it, and that it would devour quite a large invading army.*
>
> Winston Churchill
> *War in the Streets*

7-29. The urban operational framework—assess, shape, dominate, and transition—provides structure to developing considerations for defensive operations. The considerations can vary depending on the level of war at which the operation is conducted, the type of defense, and the situation. Most issues discussed may, in the right circumstances, apply to both commanders conducting major UO and commanders at lower tactical levels of command.

ASSESS

7-30. The commander defending in the urban area assesses many factors. His mission statement and guidance from higher commanders focus his assessment. If the mission is to deny a threat access to port facilities in an urban area, the commander's assessment will focus much differently than if the mission is to deny the threat control over the entire urban area. The METT-TC—mission, enemy, terrain and weather, troops and support available, time available, civil considerations—structure guides the commander's assessment. Of these, the impacts of the threat and environment—to include the terrain, weather, and civil considerations—are significant to the commander considering urban defensive operations.

The Threat

7-31. In the urban defense, a key element is the commander's assessment of the threat. One of his primary concerns is to determine the attacker's general scheme, methodology, or concept. Overall, the attacker may take one of two approaches. The most obvious would be a direct approach aimed at seizing the objectives in the area by a frontal attack. A more sophisticated approach would be indirect and begin by isolating Army forces defending the urban area. Innumerable combinations of these two extremes exist, but the threat's intentions toward the urban area will favor one approach over another. The defending Army commander (whose AO includes but is not limited to the urban area) conducts defensive planning, particularly his allocation of forces, based on this initial assessment of threat intentions. This assessment determines whether the commander's primary concern is preventing isolation by defeating threat efforts outside the area or defeating a threat attacking the urban area directly. For the higher commander, this assessment determines how he allocates forces in and outside the urban area. For the commander in the urban area, this assessment clarifies threats to sustainment operations and helps shape how he arrays his forces.

The Environment's Defensive Characteristics

7-32. A second key assessment is the defensive qualities of the urban environment. This assessment, as in any defensive scenario, is based on mission requirements and on a systemic analysis of the terrain in terms of observation and fields of fire, avenues of approach, key terrain, obstacles, and cover and concealment (OAKOC). This assessment accounts for the unique characteristics of urban terrain, population, and infrastructure as discussed in Chapter 2.

SHAPE

7-33. Commanders of a major operation shape the urban battle according to the type of defense they are attempting to conduct. If conducting an area defense or retrograde, they use shaping actions like those for any defensive action. Important shaping actions that apply to all defensive UO include—

Separating Attacking Forces from Supporting Resources

7-36. Commanders of the major operation primarily use fires and IO for separating in space and time threat forces attacking the urban area from echelons and resources in support. The purpose of this shaping action is the same as for any conventional area defense. It aims to allow the defending forces to defeat the threat piecemeal as they arrive in the urban area without support and already disrupted by deep fires and IO against information systems. This separation and disruption of the threat also sets the conditions for a mobile defense if commanders choose to execute that type of defense. These operations also prevent the threat commander from synchronizing and massing his combat power at the decisive point in the close battle.

7-37. If the urban area is part of a major mobile defense operation, the urban defense becomes the fixing force. Commanders shape the defense to encourage the threat to attack into the urban area. They lure the threat using a combination of techniques depending on the situation. They may make the urban area appear only lightly defended while other alternative courses of action appear strongly defended by friendly forces. Placing the bulk of the defending forces in concealed positions well within the urban area and positioning security forces on the periphery of the urban area portray a weak defense. In other situations, the opposite is true. If the urban area is an important objective to the threat, friendly forces can make the urban area appear heavily defended, thus ensuring that he commits sufficient combat power to the urban area to overwhelm the defense. Both cases have the same objective: to cause a major commitment of threat forces in the urban area. Once this commitment is made, the mobile defense striking force attacks and defeats the threat outside the urban area. This isolates the threat in the urban area and facilitates its destruction.

7-38. In the urban tactical battle, many shaping actions mirror those in all defensive operations. The size and complexity of the urban area prevents defending forces from being strong everywhere; shaping operations designed to engage the threat on terms advantageous to the defense have particular importance. Shaping actions include reconnaissance and security operations, passages of lines, and movement of reserve forces prior to their commitment. In addition, shaping operations critical to urban defense include mobility and countermobility operations, offensive IO, economy of force operations, and population management operations.

Creating a Mobility Advantage

7-39. In urban terrain, countermobility operations can greatly influence bringing the threat into the engagement areas of defending forces. Countermobility operations—based on understanding the urban transportation system, design, and construction characteristics—can be unusually effective (see Chapter 2). Demolitions can have important implications for creating impassable obstacles in urban canyons as well as for clearing fields of fire where necessary. Careful engineer planning can make the already constrictive terrain virtually impassable to mounted forces where appropriate, thus denying the threat combined arms capabilities. Countermobility operations in urban terrain drastically increase the defense's ability to shape the attacker's approach and to increase the combat power ratios in favor of the defense. As with all aspects of UO, countermobility considers collateral damage and the second- and third-order effects of obstacle construction.

7-40. Well-conceived mobility operations in urban terrain can provide defending forces mobility superiority over attacking forces. This is achieved by carefully selecting routes between primary, alternate, and subsequent positions, and for moving reserves and counterattack forces. These routes are reconnoitered, cleared, and marked before the operation. They maximize the cover and concealment characteristics of the terrain. Using demolitions, lanes, and innovative obstacles denies the defense of these same routes.

Applying Economy of Force Measures

7-41. Economy of force is extremely important to effective tactical urban defense. A megalopolis is too large and too easily accessible for defending forces to be strong everywhere. Forces used effectively in an economy of force role enable the defending force to mass effects at decisive points. Forces used in an economy of force role execute security missions and take advantage of obstacles, mobility, and firepower to portray greater combat power than they actually possess. They prevent the threat from determining the actual disposition and strength of the friendly defense. If, contrary to expectations, they are strongly attacked, their mobility—stemming from a mounted maneuver capability, planning, and an intimate knowledge of the terrain—allows them to delay until reserves can meet the threat. Security forces in an economy of force role take position in parts of the urban area where the threat is less likely to attack.

Defensive Combat Power
Suez – October 1973

At the end of October, the Israeli Army was in the midst of effective counterattack against the Egyptian Army. The Israelis had success attacking west across the Suez Canal. Their armored divisions were attempting to achieve several objectives, to include destroying Egyptian air defense sites and completing the encirclement of the Egyptian 3rd Army, which was trapped on the canal's east side.

To completely encircle the Egyptian 3rd Army, the Israelis had to seize all possible crossing sites to it from the canal's west bank and the Red Sea. Also, as international negotiations towards a cease-fire progressed, the Israeli government wanted to capture as much Egyptian territory as possible to improve their negotiating position after hostilities.

Consequently, the Israeli Adan Armored Division was tasked to seize the Egyptian Red Sea port of Suez on the morning of 24 October. A cease-fire was to begin at 0700, and the Israeli intent was to be decisively engaged in the city by that time and then consolidate their position as part of the cease-fire compliance.

The Adan Division plan to seize Suez was a two-part operation. Each of the division's armored brigades would have a role. The 460th Brigade would attack west of the city and complete the city's encirclement. Simultaneously, the 217th Brigade would attack in columns of battalions through the city to seize three key intersections in the city. This was in accordance with standard Israeli armored doctrine for fighting in an urban area. The 217th Brigade would seize its objectives through speed, firepower, and shock action. Once the objectives were seized, infantry and armored teams would continue attacking from the secured objectives to mop up and destroy pockets of resistance. The Israeli commanders expected to demoralize the defending Egyptians—two infantry battalions and one antitank company—by this rapid attack. The armored division commander was specifically advised by his commander to avoid a "Stalingrad" situation.

The attack got off to an ominous beginning as mist greatly inhibited a scheduled aerial bombardment in support of the attack. The 217th Brigade began its attack without infantry support and was quickly stopped by antitank missiles and antitank fire. Infantry was quickly integrated into the brigade and the attack resumed.

At the first objective, the Israelis encountered their first problems. A withering barrage of small arms, antitank missiles, and antitank fire hit the lead tank battalion, including direct fire from SU-23 anti-aircraft guns. Virtually all the officers and tank commanders in the tank battalion were killed or wounded, and several tanks were destroyed. Disabled vehicles blocked portions of the road, and vehicles that turned on to secondary roads were ambushed and destroyed. The battalion, however, successfully fought its way through the first brigade objective and on to the final brigade objective.

Hastily attached paratroop infantry in company strength were next in column following the tanks. They were traveling in buses and trucks. As the lead tank battalion took fire, the paratroopers dismounted, and attempted to secure adjacent buildings. The tank battalion's action of fighting through the objective caused the paratroopers to mount up and also attempt to move through the objective. Because of their soft skinned vehicles the paratroopers were unable to remain mounted and again dismounted, assaulted, and secured several buildings that they could defend. Once inside the buildings, the paratroopers found themselves cut off, pinned down, and unable to evacuate their considerable casualties, which included the battalion commander. The paratroopers were on the initial brigade objective but were unable to maneuver and were taking casualties.

A second paratroop company also dismounted and quickly became stalled in house-to-house fighting. The brigade reconnaissance company in M113 personnel carriers brought up the rear of the brigade column and lost several vehicles and was also unable to advance.

By 1100 the Israeli attack culminated. Elements of the 217th Brigade were on all three of the brigade's objectives in the city. However, the armored battalion, which had achieved the deepest penetration, was without infantry support and under severe antitank fire. Both paratroop companies were isolated and pinned down. In addition, an attempt to link up with the paratroopers had failed. At the same time, the civilian population of the city began to react. They erected impromptu barriers, ambushed isolated Israeli troops, and carried supplies and information to Egyptian forces.

The Israeli division commander ordered the brigade to break contact and fight its way out of the city. The armored battalion was able to fight its way out in daylight. The paratroop companies were forced to wait until darkness and then infiltrated out of the city carrying their wounded with them. Israeli casualties totaled 88 killed and hundreds wounded in addition to 28 combat vehicles destroyed. Egyptian casualties were unknown but not believed to be significant.

The fight for Suez effectively demonstrates numerous urban defensive techniques. It also vividly demonstrates the significant effect on defensive combat power of the urban environment.

The Egyptian defense demonstrates how the compartmented urban terrain restricts the mobility and the massing of firepower of armored forces. Trapped in column on the road, the Israelis were unable to mass fire on particular targets nor effectively synchronize and coordinate their fires. The short-range engagement, also a characteristic of urban combat, reduced the Israeli armor protection and eliminated the Israeli armor's ability to keep out of small arms range. Thus, hand held antiarmor weapons were more effective in an urban area. Additionally, Egyptian small arms and sniper fire critically affected Israeli C2 by successfully targeting leaders.

The Egyptian defenders effectively isolated the mounted Israelis by defending and planning engagement areas in depth. The Egyptians synchronized so that they engaged the entire Israeli force simultaneously. This forced the Israelis to fight in multiple directions. It also separated the Israeli infantry from the armor and prevented the formation of combined arms teams necessary for effective urban offensive operations.

Suez also demonstrated how civilians come to the advantage of the defense. After the battle was joined, the population—by threatening isolated pockets of Israelis and building barricades—helped prevent the Israelis from reorganizing while in contact and hindered the Israelis breaking contact. The population was also a valuable source of intelligence for the Egyptians and precluded covert Israeli movement in daytime.

Suez shows the ability of a well-placed defense in depth to fix a superior force in an urban area. Despite the Israeli commander's caution to avoid a "Stalingrad," the Israeli division, brigade, and battalion commanders were quickly trapped and unable to easily break contact. Even a successful defense on the parameter of the city would not have been nearly as effective, as the Israelis would have easily broken contact once the strength of the defense was recognized.

Another key to the success of the Egyptian defense was the Israelis' inadequate reconnaissance. While the Israelis knew the approximate size of the defending forces, they had no idea of the Egyptian dispositions. In this case, time prevented adequate reconnaissance. Key to a successful defense is adequate security to obscure defense dispositions, which permits surprise and shock effect.

The Suez defense was a decisive defeat of elite Israeli forces by regular infantry units inferior in training, morale, and numbers. Total disaster was averted only because of the professionalism of the Israeli armored forces and paratroopers that permitted them to continue to fight and eventually exfiltrate the urban trap. The Israeli forces thus escaped total destruction. Suez strongly demonstrates how the enhancing effects of the urban environment on defensive combat power are significant enough to permit inferior regular forces to defeat elite formations. Since the 1973 Suez battle, US forces in Mogadishu, Somalia, and Russian forces in Grozny, Chechnya have faced similar urban defensive ambushes.

Effectively Managing the Urban Population

7-42. Another way to shape the urban defensive battle is population management. In most cases, defending force commanders are in the urban area before combat. This time gives them the chance to manage civilians. Consequently, they can better manage and protect the population (a legal requirement) and gain more freedom of action for his forces.

7-43. Managing the civilians during the defense is a function of the size, disposition, and needs of the population and the resources available to the commander. Requesting higher support or coordinating with nongovernmental organizations, private voluntary organizations, and the local civil leadership for support may make up shortages of resources. Resources devoted to population management are carefully weighed against availability, military mission requirements, and possible collateral damage affecting tactical, operational, or strategic success. It may prove impractical to evacuate an urban area's population; still, commanders attempt to create and move most civilians to protected areas. Moving the population allows defending forces to more liberally apply fires, emplace obstacles, and relieve combat units and support units of requirements to continue life support for civilians while executing combat operations. Overall, effective civil-military operations can turn a friendly (or a neutral) population into an effective force multiplier providing support to every battlefield operating system.

Planning Counterattacks

7-44. Counterattacks are also an important tool in shaping the battlefield for defensive success. Counterattacks as a shaping tool have two applications: retaining the initiative and separating forces. However, opportunity for effective counterattacks will be brief and, therefore, timing will be critical. If conducted too soon, the counterattack may expend resources required later; if conducted too late, it may not be effective. Commanders understand the effect of the urban environment on time-distance relationships; otherwise, the timing of the attack may be upset and the operation desynchronized. Additionally, commanders develop plans beyond the counterattack to exploit potential success.

DOMINATE

7-45. Dominating the urban area in a defensive operation requires decisively defeating the threat's attacks. Defensive forces use the terrain to their advantage, employ precision supporting fires, and use direct fire from protected positions aligned against carefully selected avenues of approaches and kill zones. The combat power of the defense augmented by shaping actions and the characteristics of urban terrain force culmination of the threat attack. Like urban offensive operations, domination in urban defensive operations typically results from successful actions at the tactical level of war. These actions include—

Performing Aggressive ISR

7-46. ISR efforts of the defender are focused initially on identifying relevant information about the location and nature of the threat's main effort. Once identified, the defender's ISR focus shifts to assessing the rate at which the threat attack moves to its culminating point. Indicators of culmination may be physical fatigue of soldiers, a breakdown in C2 capability, difficulty providing logistics support, or the increasing time required to reorganize small units to attack. When that culmination is achieved, friendly forces counterattack before the threat has a chance to transition to a hasty defense.

Creating Depth

7-47. Depth in the defense is the key to forcing the threat to culminate. The urban defense cannot allow itself to be penetrated nor permit forward elements to be destroyed. The defense is designed with the greatest depth possible. Defending forces weaken the threat to the fullest extent possible by attack from each position but not permit themselves to be destroyed by fires or close assault. Instead, as threat combat power builds up against individual positions, the use of mission orders permits subordinate leaders to disengage on their own initiative and move on preplanned routes to subsequent positions. Positions are designed to be mutually supporting—withdrawing from one position to a subsequent one while supporting positions cover by fires. The attacker is constantly forced to deploy and reorganize without being able to achieve decisive effects against the defender.

Executing an Effective Obstacle Plan

7-48. Obstacles in the urban defense are designed to break up the threat's combined arms capability. Separating dismounted forces from mounted forces disrupts the cohesion of the attacker and reduces his combat power. It also exposes his individual elements to the effects of asymmetric counterattack. The leading threat dismounted force can be effectively counterattacked by a friendly combined arms element while the threat armored force in the remains vulnerable to antiarmor attack by dismounted forces.

Conducting Coordinated Counterattacks

7-49. The counterattack is one of the key actions of the urban defense. However, the commanders do not counterattack unless there is a reasonable chance of success. As the attacker moves into the depth of the urban area, his forces become fatigued, attrited, and increasingly disorganized. He likely also creates an increasingly long and exposed flank. At all levels, forces defending in urban terrain look for opportunities to counterattack. As the offensive force reaches the culmination point where it can no longer continue to attack with the available forces, the defensive commander executes a preplanned and co-ordinated counterattack. The counterattack aims to regain the initiative and to make the threat fight in multiple directions. Infiltration using superior knowledge of the terrain (including intrasurface and subsurface capabilities) permits attacking the threat throughout the depth of his formations. Small-scale counterattacks focus on C2 and combat service support capabilities. These counterattacks can set the conditions for a deliberate attack leading to the ultimate destruction of the attacking threat force.

TRANSITION

7-50. Transitions in urban defensive operations occur at all levels. As with offensive operations, commanders of major operations address which units are assigned to continue to operate in the area after defensive operations have ceased. In defensive UO, this task is not as challenging as an occupation mission during urban offensive operations. The psychology of troops defending an urban area differs from those attacking into it. Defending forces become accustomed to the environment, having experience in the environment before combat. In terms of training, it is easier for follow-on missions to be assigned to a unit that has successfully defended the urban area. This course of action takes advantage of the defending units' experience in the area and its relationships with other agencies—agencies that were operating alongside the units before and possibly during the defense. In defensive operations, regardless of the civilians' attitudes, policies regarding that population are established before the successful defense, and the command likely has experience executing operations with civil authorities and other agencies. Thus, these relationships are not new nor as significant an issue as in offensive operations. Therefore, commanders are prepared to execute various stability operations and support operations or use a successful defense to springboard into more decisive offensive operations elsewhere in the commanders' AO.

Transition to Stability Operations and Support Operations

7-51. At the end of a successful urban defense, operational commanders expect civil authority, control, and jurisdiction to increase. Additionally, the civil population will be anxious to return. Defensive combat will require virtually complete military control of the urban area; however, after the successful defense, a rapid transition will occur from military control to civil or joint military and civil control afterward. This transition will require stability operations and support operations and will include tasks such as demilitarizing munitions, clearing obstacles, and searching for isolated threat pockets of resistance. Conclusion of the defensive operations also requires transition to joint civil-military tasks, such as evaluating structures for safety, restoring essential services, and possibly creating joint law enforcement. Commanders of major operations, using a civil-military operations center and G5, anticipate these requirements to ensure a smooth, successful transition.

Transition to Offensive Operations

7-52. Units that have successfully defended the urban area may then transition to offensive operations. A rapid transition to offensive operations will require identification, preparation, and training of units designated to assume missions as the defending units leave the urban area. This preparation emphasizes continuity of policies and relationships already established. A relief in place occurs. The new occupying units provide not only a continuity of policy, but also a continuity of attitude toward the urban area, its population, and its institutions.

Chapter 8

Urban Stability Operations and Support Operations

The Rangers were bound by strict rules of engagement. They were to shoot only at someone who pointed a weapon at them, but already this was unrealistic. It was clear they were being shot at, and down the street they could see Somalis with guns. But those guns were intermingled with the unarmed, including women and children. The Somalis were strange that way. Most noncombatants who heard gunshots and explosions would flee. Whenever there was a disturbance in Mogadishu, people would throng to the spot. . . . Rangers peering down their sights silently begged the gawkers to get the hell out of the way.

Black Hawk Down

The fundamental shared aims between stability operations and support operations and any operation conducted in an urban environment are the increased significance and influence of the civil population and nonmilitary organizations. Often, no military victory is to be achieved. The center of gravity for these operations normally cannot be attacked through military means alone; the Army (and the military in general) is often but one tool supporting a larger, civil-focused effort. Without a tightly coordinated civil-military effort, success will be difficult or impossible to achieve. Commanders who can understand and cope with the complexities of stability operations and support operations gain insights that directly apply to executing any urban operation. Urban stability operations and support operations may complement urban offensive and defensive operations, or may dominate the overall operation. Army forces may need to conduct offensive and defensive operations to defend themselves or destroy urban threats seeking to prevent the decisive stability or support mission. During hostilities, urban stability operations may keep armed conflict from spreading, encourage coalition and alliance partners, and secure the civilian population's support in unstable urban areas (and the surrounding rural areas under their influence). Following hostilities,

CONTENTS

urban stability operations may provide a secure environment for civil authorities to rebuild. Urban support operations can range from transporting, feeding, and sheltering the population made homeless as a result of combat operations or natural disasters to providing medical care during urban counterinsurgency operations.

PURPOSE OF URBAN STABILITY OPERATIONS AND SUPPORT OPERATIONS

8-1. Army forces conduct stability operations and support operations to deter war, resolve conflict, promote peace, strengthen democratic processes, retain United States (US) influence or access abroad, assist US civil authorities, and support moral and legal imperatives. Stability operations promote and sustain regional and global stability. In contrast, support operations meet the urgent needs of designated groups, for a limited time, until civil authorities can accomplish these tasks without military assistance. Nearly every urban operation will involve some type or form of stability operation or support operation combined and sequenced with offensive and defensive operations.

CHARACTERISTICS OF URBAN STABILITY OPERATIONS AND SUPPORT OPERATIONS

8-2. Worldwide urbanization, migration trends from rural to urban areas, and more centralized populations in urban areas increase the chance that Army forces will conduct stability operations and support operations in or near urban areas. Simply put, many people live in urban areas, and their welfare will be the primary reason for conducting these operations. Urban areas that serve as economic and government centers (the ideal location for US and allied embassies) are often the focal point for many threat activities. Therefore, Army forces may need to conduct stability operations in these cities to counter those threats. Additionally, urban areas may contain the resources and infrastructure to support both types of operations, regardless of whether the overall focus is in urban or rural areas. Repairing or restoring the infrastructure may be a critical task in accomplishing a support mission. Supported governmental and nongovernmental agencies are not as logistically self-sufficient as the Army. As such, these agencies may need to center their operations in and around urban areas to use the area's infrastructure to support themselves and their objectives. These agencies may require military protection to accomplish their missions. Figure 8-1 lists some defining characteristics of these wide-ranging operations.

Figure 8-1. Characteristics of Stability Operations and Support Operations

8-3. Stability operations and support operations are diverse, varied in duration, unilateral or multinational, and domestic or foreign. Like all urban operations (UO), they are usually joint. Unlike urban offensive and defensive operations, they are more often interagency operations and require more restrictive rules of engagement (ROE). The multiplicity of actors involved usually increases the scope and scale of required coordination and communication. In urban stability operations or support operations, adverse conditions arising from natural or man-made disasters or other endemic conditions—such as human suffering, disease, violations of human rights, or privation—will significantly modify the urban environment. Unresolved political issues and tenuous agreements, difficulties discriminating combatants from noncombatants or between parties of a dispute, and the absence of basic law and order all serve to complicate an already complex and uncertain environment. Civil-military and legal considerations take on added significance in all urban operations (see Civil-Military Operations and Legal Support in Chapter 9), but even more so in urban stability operations and support operations. Finally, recognizing and achieving the desired end state is often more difficult than in offensive and defensive operations.

8-4. Overall, commanders of major operations involving urban stability operations and support operations do not expect clear guidance. They learn, adapt, and live with ambiguity. They cannot expect to operate in a political vacuum (even commanders at the tactical level) and do not expect an easily identifiable enemy located across a clearly demarcated line. In fact in many peace operations, commanders and their soldiers resist the need to have an enemy—difficult at best when one side or another (or both) may be sniping at them. They also expect changing and additional missions and tasks, without being allowed to use every means at their disposal to carry out those missions. Many tasks required may be ones for which their units have never, or rarely, trained. Finally, commanders expect to show restraint with a keen sensitivity to political considerations and to alien cultures, either or both of which they might find confusing or even repugnant.

URBAN STABILITY OPERATIONS, SUPPORT OPERATIONS, AND BATTLEFIELD ORGANIZATION

8-5. Each type of urban stability operation or support operation is distinct. These operations differ even more when applied to a specific urban area. Due to the complexity of the environment, commanders carefully arrange their forces and operations according to purpose, time, and space to accomplish the mission. In most UO the terrain, the dense population (military and civilian), and the participating organizations will further complicate this arrangement.

DECISIVE OPERATIONS

8-6. In urban stability operations, decisive operations may take many years and include multiple actions before achieving the desired end state. This particularly applies to the strategic and operational levels. Oppositely, decisive operations involved in an urban support operation for mitigating or reducing disease, hunger, privation, and the effects of disasters normally achieve faster results. However, an operation that attacks the underlying cause and seeks to prevent or relieve such conditions is more a stability operation than a support operation and will usually take longer. In urban areas, establishing law and order to protect critical infrastructure and the inhabitants from lawlessness and violence is often critical and often the decisive operation.

SHAPING OPERATIONS

8-7. Shaping operations establish and maintain the conditions for executing decisive operations. In urban stability operations and support operations, shaping operations always include information operations (IO) that influence perceptions and maintain legitimacy. Often, various participants, and their potentially divergent missions and methods, are involved. Army commanders coordinate their planning and efforts (early and continuously) to ensure that their decisive, shaping, or sustaining operations are not working against other agencies' efforts and operations—agencies that may have the lead role in the operation. Thus, a critical shaping operation may be to establish the coordination to help develop a common purpose and direction among agencies. In some instances and with some organizations and agencies, particularly nongovernmental organizations (NGOs), genuine unity of effort may not be achievable; however, recognizing the differences in aims and goals will allow Army commanders to conduct operations with less friction. Commanders include NGOs and appropriate governmental agencies in mission readiness exercises or any other training for stability operations or support operations.

SUSTAINING OPERATIONS

8-8. Sustaining operations enable decisive and shaping operations and include combat service support, rear area and base security, movement control, terrain management, and infrastructure development. Sustainment bases, especially those located in urban areas, become an attractive target for hostile civilians; therefore, commanders actively and aggressively protect these bases as well as lines of communications (see Chapter 9).

TYPES AND FORMS OF STABILITY OPERATIONS AND SUPPORT OPERATIONS

8-9. Figure 8-2 defines stability operations and support operations and lists their subordinate types and forms. If necessary, commanders can refer to FM 3-07 to develop a more detailed understanding of the specifics of these diverse operations.

STABILITY OPERATIONS

8-10. Urban areas will be decisive to accomplishing many types of stability operations because urban areas are the centers of population, culture, economy, and government. Much of the support provided by Army forces will aim to assist local, regional, or national governments. Their location (urban areas) will, by necessity, be a dominating factor in accomplishing the mission. As importantly, many stability operations—enforcing peace in Bosnia for example—will require interacting, influencing, controlling, or protecting all or parts of the civilian population. Assessing, understanding, and gaining the support of civilians in key economic, cultural, or political urban areas may influence surrounding regions (smaller urban areas and the rural country-side) and may be decisive to achieving stability objectives. Finally, the support and assistance that Army forces will provide is only temporary although often of long duration. Commanders execute operations with that thought always in mind. Eventually, the government and administration secure and support their population by themselves.

	Definition	Types or Forms
Stability	Operations that promote and protect US national interests by influencing the threat, political, and information dimensions of the operational environment through a combination of peacetime developmental, cooperative activities and coercive actions in response to crisis.	• Peace Operations • Foreign Internal Defense • Security Assistance • Support to Insurgencies • Humanitarian and Civic Assistance • Support to Counterdrug Operations • Combatting Terrorism • Noncombatant Evacuation Operations • Arms Control • Show of Force
Support	Operations that employ Army forces to assist civil authorities, foreign or domestic, as they prepare for or respond to crisis and relieve suffering.	• Domestic Support Operations • Foreign Humanitarian Assistance Forms • Relief Operations • Support to Domestic Chemical, Biological, Radiological, Nuclear, and High-Yield Explosive Consequence Management • Support to Civil Law Enforcement • Community Assistance

Figure 8-2. Urban Stability Operations and Support Operations

SUPPORT OPERATIONS

8-11. Support operations consist of domestic support and foreign humanitarian assistance operations. They can occur in a foreign urban environment as a result of military operations affecting the infrastructure or from a natural disaster, such as an earthquake. Support operations can also occur domestically when a natural or man-made emergency overwhelms local resources. Such a situation could result from a chemical, biological, radiological, nuclear, and high-yield explosive (CBRNE) incident, hurricane, flood, or civil disturbance in a domestic urban area.

8-12. CBRNE incidents are disastrous. In urban areas, the potential for catastrophic loss of life and property is enormous. The Army categorizes CBRNE incidents separately from other natural and man-made disasters because it has specific expertise with these weapons. CBRNE incidents usually result from a military or terrorist threat (adding a law enforcement dimension to the disaster).

8-13. Subways and other subsurface areas offer ideal areas for limited chemical or biological attacks. Nuclear attack (and high-yield explosives) can produce tragic results due to the effects of collapsing structures, flying debris, and fires. Dispersion patterns are affected by the urban terrain and are more difficult to predict and monitor. Large-scale incidents may produce hundreds of thousands of casualties, but even a limited attack may require evacuating and screening large numbers of civilians. Requirements for medical support, basic life support, and, if necessary, decontamination may quickly overwhelm the Army force's capabilities even with augmentation.

8-14. Panic and disorder may accompany the event. Fleeing civilians may clog elements of the transportation and distribution infrastructure. Physical destruction may also affect other components of the infrastructure of critical and immediate concern, such as energy and administration and human services (water, sanitation, medical, fire fighting, and law enforcement). Because all elements of the infrastructure may be affected, the overall recovery time may be lengthened and the effects broadened to include much of the surrounding area. The effects of a single urban CBRNE event potentially could be felt nationally or globally.

CONSIDERATIONS OF URBAN STABILITY OPERATIONS AND SUPPORT OPERATIONS

> *In wars of intervention the essentials are to secure a general who is both a statesman and a soldier; to have clear stipulations with the allies as to the part to be taken by each in the principal operations; finally, to agree upon an objective point which shall be in harmony with the common interests.*
>
> Lieutenant General Antoine-Henri, Baron de Jomini

8-15. The urban operational framework (assess, shape, dominate, and transition) provides a structure for developing considerations unique to urban stability operations and support operations. Many considerations presented in urban offensive and defensive operations apply to urban stability operations and support operations, particularly those that address how to assess the urban and overall operational environment. Because the situations in which stability operations and support operations normally occur share strong similarities with any urban environment, many of these considerations are closely linked to the urban fundamentals presented in Chapter 5. Taken together, commanders will often find them useful in conducting UO throughout the full range of operations and across the spectrum of conflict. Appendix C has a historical example of how to apply the urban operational framework to support operations and stability operations.

ASSESS

8-16. In urban stability operations and support operations, commanders carefully assess the political dimension of the operational environment, as well as their role and the media's part in managing information. These operations are inherently tied to the exercise of diplomatic power. All operations in urban areas are often the focus of the media and thus gain considerable public and political attention. Therefore, military objectives in urban stability operations and support operations are more directly linked with political objectives. The relationship between the levels of war—strategic, operational, and tactical—is often closer than in urban offensive and defensive operations. Military objectives are carefully nested within political objectives. Commanders ensure that the ways and means to accomplish their objectives, to include security and force protection measures, will hold up to media scrutiny and are appropriate for the situation and environment. All levels of command understand the link between political and military objectives, to include a basic understanding at the individual level. One uncoordinated, undisciplined, or inappropriate action, even at the lowest level, could negate months or years of previous, disciplined effort. Commanders balance security and force protection measures with mission accomplishment. Ineffective measures can put soldiers at too great a risk and jeopardize the mission. Conversely, overly stringent measures may make it difficult for forces to interact with the population closely—essential in many of these operations. Finally, commanders will need a thorough assessment of the governmental and nongovernmental organizations and agencies that will be operating in or near urban areas that fall within their area of operations.

Political and Military Objectives

> Commanders consult the US Agency for International Development's (USAID) *Field Operations Guide for Disaster Assessment and Response* when conducting their assessments and developing measures of effectiveness for many urban relief operations.

8-17. Commanders translate political objectives into military objectives that are clear and achievable (clear tasks and purposes) and can lead to the desired end state. Political objectives may be vague making it difficult for commanders to conduct their mission analysis. This applies to tactical- and even operational-level commanders, unskilled at higher level, strategic political-military assessments. Each type of stability operation or support operation is distinct, often unfamiliar to the executing unit, and unique to the specific situation. These factors often make it difficult to confidently determine the specific tasks that will lead to mission success. Therefore, commanders also establish measures of effectiveness that aid in understanding and measuring progress and help gauge mission accomplishment.

8-18. These criteria should be measurable (in some circumstances, a qualitative assessment may be most appropriate) and link cause with effect. They help determine the changes required and are essential to the assessment cycle required for urban stability operations and support operations. In a humanitarian relief operation to aid the starving, commanders could determine that the decisive effort is delivering safe food to the urban area. To judge success or effectiveness, they could determine that the measure is the number of food trucks dispatched daily to each distribution site; the more trucks, the more effective the efforts. However, this measure must correlate with the overarching measure of effectiveness: decline in the mortality rate. If no significant decrease in deaths due to starvation occurs, they may need to reassess and modify the tasks or measure of effectiveness. A better measure may be to track the amount of food consumed by those in need instead of simply counting the number of trucks dispatched.

8-19. However, planners also be wary of the unintended consequences of well-intentioned urban support operations. For example, providing free, safe food may alleviate starvation, but could also undercut the local agricultural system by reducing demand in the market. If the food is distributed through urban centers, urbanization could increase, further reducing the food supply and adding to the existing strains on the infrastructure. Areas around which measures of effectiveness can be formed for many stability operations and support operations (including the example above) and which will help return most societies to some degree of normalcy and self-sufficiency include:

establishing force protection for Army forces operating in the urban environment. However, many such operations, particularly stability operations, require extra time to forge a lasting change. Over time, and particularly in peacetime when objectives center on helping others and avoiding violence, even the complex urban environment may seem benign. Without continued, aggressive command emphasis, soldiers may become lulled into complacency. It is usually then that Army forces are most vulnerable to terrorist tactics, such as bombings, kidnappings, ambushes, raids, and other forms of urban violence.

8-22. Although force protection will not ensure successful urban stability operations or support operations, improper assessment and inadequate force protection measures can cause the operation to fail. In either operation, keeping a neutral attitude toward all elements of the urban population, while maintaining the appropriate defensive posture, enhances security. One threat principle discussed in Chapter 3 was that threats would seek to cause politically unacceptable casualties. An improper threat assessment and a lapse in security at the tactical level could result in casualties. That result could affect strategy by influencing domestic popular support and subsequently national leadership decisions and policy.

8-23. Emphasizing security and force protection measures does not mean isolating soldiers from contact with the urban population. On the contrary, commanders balance survivability with mobility according to the factors of METT-TC—mission, enemy, terrain and weather, troops and support available, time available, civil considerations. Survivability measures—such as hardening or fortifying buildings and installations, particularly where large numbers of soldiers are billeted— may be required. On the other hand, mobility operations are essential in preserving freedom of action and denying a threat the opportunity to observe, plan, and attack urban forces. Mission degradation and increased risk to the force can result if force protection measures prevent Army forces from conducting prudent missions and establishing an active and capable presence.

Assessment of Security and Force Protection
Belfast, Northern Ireland

Since 1969, Belfast has significantly affected the British military campaign for stabilizing the area. British operations in Belfast illustrate the difficulty of balancing the security and protecting forces with maintaining the stabilizing presence necessary to uphold law and order, minimize violence, and control the urban population.

British successes in protecting Belfast's infrastructure and government facilities from terrorist attacks compelled various terrorist cells, especially the Irish Republican Army and the Provisional Irish Republican Army, to attack more military targets. At the time, British soldiers and bases presented relatively unprotected targets to these factions, and attacks against them solidified their legitimacy as an "army." In response, British commanders implemented extreme security and force protection measures—from ballistic protection vests and helmets to fortress-like operational bases and large unit patrols. These protection measures successfully decreased the violence against British soldiers in Belfast. However, they also decreased the soldiers' interaction with the population and their ability to stabilize the city. The large patrols, while protecting the soldiers, inhibited effective saturation of neighborhoods. These patrols, coupled with fortress-like bases and bulky protective clothing, created an "us-versus-them" mentality among civilians and soldiers.

As force protection increased and stabilizing effects decreased, the terrorists were provided more targets of opportunity among the civilians and infrastructure. British commanders reassessed the situation, identified this "see-saw" effect, and adapted to strike a better balance between force protection and effective presence patrols. For example, British forces switched to four-man patrols to enable greater mobility and wore berets instead of helmets to appear less aggressive.

Participating Organizations and Agencies

8-24. Across the spectrum of urban operations, but more so in these operations, numerous NGOs may be involved in relieving adverse humanitarian conditions. Dense populations and infrastructure make an urban area a likely headquarters location for them. In 1994 during OPERATION UPHOLD DEMOCRACY, for example, over 400 civilian agencies and relief organizations were operating in Haiti. Therefore, commanders assess all significant NGOs and governmental agencies operating (or likely to operate) in or near the urban area to include their—

SHAPE

8-25. Commanders conduct many activities to shape the conditions for successful decisive operations. In urban stability operations and support operations, two rise to the forefront of importance: aggressive IO and security operations.

Vigorous Information Operations

8-26. IO, particularly psychological operations (PSYOP) and the related activities of civil affairs (CA) and public affairs, are essential to shape the urban environment for the successful conduct of stability operations and support operations. Vigorous IO can influence the perceptions, decisions, and will of the threat, the urban population, and other groups in support of the commander's mission. IO objectives are translated to IO tasks that are then executed to create the commander's desired effects in shaping the battlefield. These operations can isolate an urban threat from his sources of support; neutralize hostile urban populations or gain the support of neutral populations; and mitigate the effects of threat IO, misinformation, rumors, confusion, and apprehension.

Security Operations

8-27. **Protecting Civilians.** Security for NGOs and civilians may also be an important shaping operation, particularly for support operations. Commanders may need to provide security to civil agencies and NGOs located near or operating in the urban area so that these agencies can focus their relief efforts directly to the emergency. Commanders may also need to protect the urban population and infrastructure to maintain law and order if the urban area's security or police forces are nonexistent or incapacitated.

8-28. **Preserving Resources.** Just as forces are at risk during urban stability operations or support operations, so are their resources. In urban areas of great need, supplies and equipment are extremely valuable. Criminal elements, insurgent forces, and people in need may try to steal weapons, ammunition, food, construction material, medical supplies, and fuel. Protecting these resources may become a critical shaping operation. Otherwise, Army forces and supporting agencies may lack the resources to accomplish their primary objectives or overall mission.

Prioritize Resources and Efforts

8-29. Urban commanders will always face limited resources with which to shape the battlefield, conduct their decisive operations, and accomplish their objectives. They prioritize, allocate, and apply those resources to achieve the desired end state. Especially in urban support operations, they tailor their objectives and shape their operations to achieve the greatest good for the largest number. Commanders first apply the urban fundamental of preserving critical infrastructure to reduce the disruption to the residents' health and welfare. Second, they apply the urban fundamental of restoring essential services, which includes prioritizing their efforts to provide vital services for the greatest number of inhabitants possible.

DOMINATE

> *If there is any lesson to be derived from the work of the regular troops in San Francisco, it is that nothing can take the place of training and discipline, and that self-control and patience are as important as courage.*
>
> Brigadier General Frederick Funston commenting on the Army's assistance following the 1906 San Francisco earthquake and fire

8-30. The focus of the Army is warfighting. Therefore, when Army commanders conduct many urban stability operations and support operations, they adjust their concept of what it means to dominate. Commanders will most often find themselves in a supporting role and less often responsible for conducting the decisive operations. They accept this supporting function and capitalize on the professional values instilled in each soldier, particularly the sense of duty to do what needs to be done despite difficulty, danger, and personal hardship. Commanders also put accomplishing the overall mission ahead of individual desires to take the lead in Chapter 5 to encourage and allow subordinates to exercise creative and critical thinking required for planning and executing these UO.

Aggressive Coordination

8-33. In urban stability operations and support operations, the increased number of participants (both military and nonmilitary) and divergent missions and methods create a significant coordination challenge. Significant potential for duplicated effort and working at cross-purposes exists. The success of UO often depends on establishing a successful working relationship with all groups operating in the urban area. The absence of unity of command among civil and military organizations does not prevent commanders from influencing other participants not under his direct command through persuasion, good leadership, and innovative ideas.

Support of and Coordination with Civilian Authorities: The 1992 Los Angeles Riots

During the spring of 1992, soldiers from the 40th Infantry Division, California National Guard were among the forces deployed to Los Angeles County to assist the California Highway Patrol, Los Angeles County Sheriffs, and civilian law enforcement. They worked to quell the riots that were sparked by the "not guilty" verdicts concerning four police officers who, following a lengthy high-speed chase through Los Angeles, were accused of brutally beating Rodney King.

Successful accomplishment of this support operation was attributed to the exercise of strong Army leadership and judgment at lower tactical levels, particularly among the unit's noncommissioned officers. An essential component of combat power, it was especially critical in executing noncontiguous and decentralized operations in the compartmented terrain of Los Angeles. As important, however, was the clear understanding that Army forces were to support civilian law enforcement—and not the other way around. The 40th Infantry Division aligned its area of operations with local law enforcement boundaries and relied heavily on police recommendations for the level at which soldiers be armed (the need for magazines to be locked in weapons or rounds chambered).

One incident emphasized the need for coordination of command and control measures with civilian agencies even at the lowest tactical levels. To civilian law enforcement and Army forces, the command "Cover me" was interpreted the same: be prepared to shoot if necessary. However, when a police officer responding to a complaint of domestic abuse issued that command to an accompanying squad of Marines, they responded by immediately providing a supporting base of fire that narrowly missed some children at home. However, the Marines responded as they had been trained. This command meant something entirely different to them than for Army soldiers and civilian law enforcement. Again, coordination at all levels is critical to the success of the operation (see also the vignette in Appendix B).

8-34. In the constraints imposed by METT-TC and operations security (OPSEC), commanders seek to coordinate all tactical stability operations with other agencies and forces that share the urban environment. Commanders strive to overcome difficulties, such as mutual suspicion, different values and motivations, and varying methods of organization and execution. Frequently, they *initiate* cooperative efforts with participating civilian agencies and determine where their objectives and plans complement or conflict with those agencies. Commanders then match Army force capabilities to the needs of the supported agencies. In situations leading to many urban support operations, confusion may initially make it difficult to ascertain specific priority requirements. Reconnaissance and liaison elements—heavily weighted with CA and health support personnel—may need to be deployed first to determine what type of support Army forces provide. Overall, aggressive coordination will make unity of effort possible in urban stability operations or support operations where unity of command is difficult or impossible to achieve.

Perseverance

8-35. The society is a major factor responsible for increasing the overall duration of urban operations. This particularly applies to urban stability operations and support operations where success often depends on changing people's fundamental beliefs and subsequent actions. Modifying behavior requires influence, sometimes with coercion or control, and perseverance. They often must be convinced or persuaded to accept change. This may take as long or longer than the evolution of the conflict. Decades of problems and their consequences cannot be immediately corrected. Frequently, the affected segments of the urban society must see that change is lasting and basic problems are being effectively addressed.

8-36. In most stability operations, success will not occur unless the host nation, not Army forces, ultimately prevails. The host urban administration addresses the underlying problems or revises its policies toward the disaffected portions of the urban population. Otherwise, apparent successes will be short lived. The UO fundamental of understanding the human dimension is of paramount importance in applying this consideration. After all Army forces, particularly commanders and staff of major operations, understand the society's history and culture, they can begin to accurately identify the problem, understand root causes, and plan and execute successful Army UO.

Reasonable Restraint

8-37. Unlike offensive and defensive operations where commanders seek to apply overwhelming combat power at decisive points, restraint is more essential to success in urban stability operations and support operations. It involves employing combat power selectively, discriminately, and precisely (yet still at decisive points) in accordance with assigned

missions and prescribed legal and policy limitations. Similar to the UO fundamentals of minimizing collateral damage and preserving critical infrastructure, restraint entails restrictions on using force. Commanders of major operations issue or supplement ROE to guide the tactical application of combat power. Excessively or arbitrarily using force is never justified or tolerated by Army forces. Even unintentionally injuring or killing inhabitants and inadvertently destroying their property and infrastructure lessens legitimacy and the urban population's sympathy and support. It may even cause some inhabitants to become hostile. In urban stability operations and support operations, even force against a violent opponent is minimized. Undue force often leads to commanders applying ever-increasing force to achieve the same results.

8-38. Although restraint is essential, Army forces, primarily during urban stability operations, are always capable of limited combat operations for self-defense. This is in accordance with the UO fundamental of conducting close combat. This combat capability is present and visible, yet displayed in a nonthreatening manner. A commander's intent normally includes demonstrating strength and resolve without provoking an unintended response. Army forces are capable of moving quickly through the urban area and available on short notice. When necessary, Army forces are prepared to apply combat power rapidly, forcefully, and decisively to prevent, end, or deter urban confrontations. Keeping this deterrent viable requires readiness, constant training, and rehearsals. It also requires active reconnaissance, superb OPSEC, a combined arms team, and timely and accurate intelligence, which in the urban environment requires a well-developed HUMINT capability.

Resolute Legitimacy

8-39. Closely linked to restraint is legitimacy or the proper exercise of authority for reasonable purposes. Achieving or maintaining legitimacy during urban stability operations or support operations is essential in obtaining the support of its population. Commanders can ensure legitimacy by building consent among the population, projecting a credible force, and appropriately using that force. Perceptions play a key role in legitimacy, and skillful IO can shape perceptions. Commanders send messages that are consistent with the actions of their forces. Generally, the urban population will accept violence for proper purposes if that force is used impartially. Perceptions that force is excessive or that certain groups are being favored over others can erode legitimacy. A single soldier's misbehavior can significantly degrade a commander's ability to project an image of impartiality and legitimacy.

TRANSITION

8-40. Commanders of major operations are the focal point for synchronizing tactical stability operations and support operations with strategic diplomatic and political issues. They are also the critical links between national intelligence resources and the tactical commander. Because strategic, diplomatic, and political changes can quickly transition the type of urban operation, they keep subordinate tactical commanders abreast of changes in intelligence, policy, and higher decisions. The potential to rapidly transition to urban combat operations emphasizes the need to maintain the capability to conduct close, urban combat. Failure to recognize changes and transition points may lead to UO that do not support the attainment of the overall objective and needlessly use resources, particularly soldiers' lives. Therefore, Army forces on the ground in an urban stability operation are more aware of the strategic environment than the threat and the civilian population, each of whom will have their own means of monitoring the national and international situation.

Legitimate and Capable Civilian Control

8-41. Commanders maintain or enhance the credibility and legitimacy of the government and police of the urban area and of the host nation's military forces operating there. In accordance with the urban fundamental of transitioning control, urban commanders conclude UO quickly and successfully, often to use assets elsewhere in their area of operations. This entails returning the control of the urban area back to civilian responsibility as soon as feasible. The host nation's military and the urban area's leadership and police are integrated into all aspects of the urban stability operations or support operations to maintain their legitimacy. They are allowed (or influenced) to take the lead in developing and implementing solutions to their own problems.

8-42. If the host nation's leadership, military, and police are not up to the task, commanders can take steps to increase its capabilities through training, advice, and assistance by CA units or by other nongovernmental or governmental organizations and agencies. Sometimes, new leadership and a restructured police force may be required, particularly when corrupt and no longer trusted by the population. This candid assessment of the urban leadership's ability to govern, protect, and support itself is made early in the planning process. Only then can commanders ensure that resources and a well thought-out and coordinated plan (particularly with civilian organizations) are available for a speedy transition. IO will be paramount in these instances to ensure that the urban population sees the training and rebuilding process itself as legitimate. Throughout urban stability operations and support operations, commanders shape the conditions to successfully hand over all activities to urban civilian authorities.

Longer-Term Commitment

8-43. Many stability operations often require perseverance and a longer-term US commitment requiring a rotation of Army units into the area of operations to continue the mission. Considerations for these transitions are similar to a relief in place (see FM 3-90) combined with considerations for deployment and redeployment. FM 41-10 contains a comprehensive appendix on transition planning and coordination activities applicable to UO. The commander of the major operation ensures that the incoming unit understands the political and strategic objectives behind the tasks that they accomplish. Otherwise, the new unit may begin to plan operations that are similar to those conducted by the previous unit without achieving the desired end state or accomplishing the mission.

Chapter 9

Urban Combat Service Support

Even supply is different. While deliveries do not need to be made over great distances, soft vehicles are extremely vulnerable in an environment where it is hard to define a front line and where the enemy can repeatedly emerge in the rear. All soldiers will be fighters, and force and resource protection will be physically and psychologically draining. Urban environments can upset traditional balances between classes of supply. . . . [a] force may find itself required to feed an urban population, or to supply epidemic-control efforts. . . . [a]ll combat service support troops are more apt to find themselves shooting back during an urban battle than in any other combat environment.

Ralph Peters
"Our Soldiers, Their Cities"

Combat service support (CSS) capabilities exist to enable the Army to initiate and sustain full spectrum operations. CSS is a major component of sustaining operations and provides the means for commanders to build and maintain combat power. Sustaining operations are inseparable from decisive and shaping operations. In offensive and defensive operations, they are not by themselves likely to be decisive or shaping; however, they contribute to those operations. In some stability operations and most support operations, when the critical objectives may be restoring the infrastructure and the welfare of civilians, CSS forces can often be the decisive

CONTENTS

element. Their success will allow Army forces to *dominate* this complex environment. However, like all urban operations (UO), CSS operations affect and are affected by the environment. The urban terrain, infrastructure, and existing resources, coupled with supportive civilians, may facilitate CSS operations. In contrast, a poorly designed or damaged infrastructure and a hostile population may severely hamper CSS operations. In the latter case, critical Army resources required elsewhere in the area of operations (AO) may be diverted to repair facilities and control and support the inhabitants of the urban areas.

URBAN CSS CHARACTERISTICS

9-1. CSS characteristics (see Figure 9-1) guide prudent logistic planning regardless of the environment. They provide commanders an excellent framework to analyze and develop urban logistic requirements, assess the impact of the environment on the provision of CSS, and gauge the effectiveness of urban CSS support.

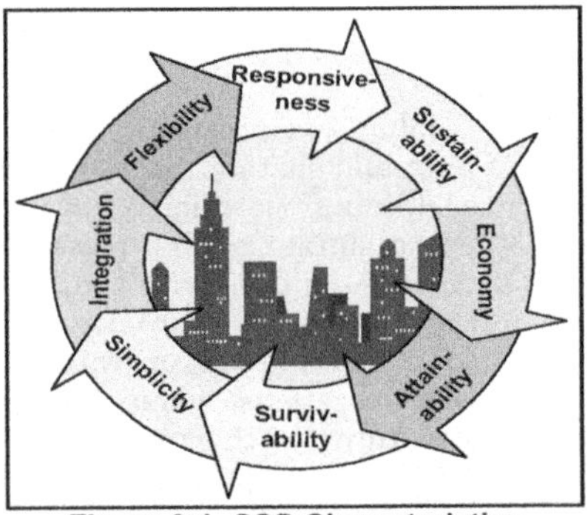

Figure 9-1. CSS Characteristics

RESPONSIVENESS AND SUSTAINABILITY

9-2. UO require responsiveness and sustainability to establish and maintain the tempo necessary for success. Responsiveness—providing the right support in the right place at the right time—is the essential CSS characteristic. It requires that CSS commanders and planners accurately forecast urban operational requirements. Continuous urban operations will drain personnel, equipment, and supplies (based on history, this can be more than five times that experienced in other environments). Therefore, sustainability—the ability to maintain continuous support throughout all phases of the operation—will be a significant concern. Anticipation is critical to both responsiveness and sustainability. It requires that CSS commanders and planners comprehend the potential effects that the components of the urban environment (terrain, infrastructure, and society) may have on operations and CSS, either benefiting or impeding UO. Effective urban operational and logistic planning cannot be accomplished separately. Operational and CSS planners, as well as CSS operators, are closely linked to aid in synchronizing and attaining responsiveness and sustainability.

ECONOMY AND ATTAINABILITY

9-3. A thoughtful assessment and understanding of the urban environment can also help determine how specific urban areas can contribute to or frustrate the achievement of economy and attainability. Economy is providing the most efficient support at the least cost to accomplish the mission. Attainability means generating the minimum essential supplies and services necessary to begin operations. If available, obtaining support in the AO costs less than purchasing the supplies outside the area and then transporting them there. Critical resources may be available in urban areas to support the operation. However, relying on sources outside the established military logistic system may create conflict with other CSS characteristics. A strike by longshoremen, for example, may shut down port operations (at least temporarily) lowering responsiveness and sustainability.

SURVIVABILITY

9-4. Survivability is being able to protect support functions from destruction or degradation. Commanders often choose to locate CSS functions in an urban area because the buildings may better protect and conceal equipment, supplies, and people. Urban industrial areas are frequently chosen as support areas because they offer this protection as well as sizeable warehouses, large parking areas, and materials handling equipment (MHE). Such areas facilitate the storage and movement of equipment and supplies. They also provide readily available water, electricity, and other potentially useful urban resources and infrastructure. However, these areas may also contain toxic industrial materials (TIM) (see the discussion of industrial areas in Chapter 2). These materials and chemicals in close proximity to support areas may unjustifiably increase the risk to survivability, especially any CSS facilities located in subsurface areas (liquids and heavier gases often sink and accumulate in low-lying areas). Furthermore, CSS activities located in any type of confined urban areas can offer lucrative targets for terrorists or even angry crowds and mobs. Although host-nation support may include assets to assist in defending CSS units and lines of communications (LOCs), CSS commanders carefully consider if adequate protection measures can ensure survivability.

> **Base Security:**
> **Tan Son Nhut, Vietnam – Tet 1968**
>
> Colonel Nam Truyen was the commander of the 9th Vietcong Division who planned and conducted the attack on the US airbase at Tan Son Nhut during the 1968 Tet Offensive. He had previously entered the airbase during the 1967 Christmas cease-fire using forged identity papers to conduct his own personal reconnaissance.

SIMPLICITY

9-5. Simplicity is required in both planning and executing CSS operations in this complex environment. Developing standard procedures among the Army, other services, and especially civilian governmental and nongovernmental agencies; of liaison and open channels of communication; between simple plans and orders; and extensive rehearsals contribute immeasurably to attaining this necessary characteristic.

INTEGRATION

9-6. The need for CSS integration increases in urban operations due to the joint nature of UO and greater numbers of other governmental and non-governmental agencies operating in or near urban areas. More nongovernmental organizations (NGOs) will likely exist because urban areas often contain most of a region's population. Most NGOs focus on people. Army forces and other military and nonmilitary groups cooperate and coordinate their actions. Much of their coordination will revolve around logistics. Cooperation and coordination will take advantage of each group's logistic capabilities, help to avoid duplicated effort (contributing to economy), and create logistic synergy. It will also help to curtail competition for the same urban resources and assist in developing a unified list of priorities. Such coordination will help ensure that other operations by one force or agency will not disrupt or destroy portions of the urban infrastructure critical to another's logistic operations and the overall mission. (See the discussion of coordination with other agencies included at the end of Chapter 4.)

FLEXIBILITY

9-7. Lastly, commanders develop flexibility. Although they and their staffs thoroughly understand the urban environment essential to planning CSS operations, they cannot anticipate every eventuality. Urban commanders possess the ability to exploit fleeting opportunities. Knowledge of the environment, particularly its infrastructure, can aid in developing innovative solutions to CSS acquisition and distribution problems. Flexibility enables CSS personnel to remain responsive to the force commander's needs.

9-8. The force and CSS commanders consider and prioritize these characteristics as they visualize UO. Each characteristic does not affect every operation and urban area in the same way. The CSS characteristics seldom exert equal influence, and their importance varies according to mission, enemy, terrain and weather, troops and support available, time available, civil considerations (METT-TC). Like the principles of war, commanders do not ignore the potential impact of CSS characteristics and how their influence changes as the operation evolves (see FM 100-10).

LOGISTICS PREPARATION OF THE THEATER

9-9. A thorough logistics preparation of the theater (LPT) is critical for an adaptable UO logistic support plan. CSS planners conduct the LPT to *assess* the situation from a logistic perspective and determine how best to support the force commander's plan. CSS planners understand the urban environment, the fundamentals of UO, and the urban environment's effects on combat service support (as well as the other battlefield operating systems). Such knowledge allows the planners to develop a detailed estimate of support requirements. A thorough LPT helps commanders determine the most effective method of providing adequate, responsive support to meet support estimates while minimizing the CSS footprint. Overall, it helps tie together UO requirements with acquisition and distribution. As with all operations, but particularly in a dynamic urban environment, this assessment process is continuous since requirements will change as the urban operation unfolds and matures.

SUPPORT TO IPB

> *CSS planning accounts for increased consumption, increased threats to lines of communications, and anticipated support to noncombatants. . . . Urban operations place a premium on closely coordinated, combined arms teams and carefully protected CSS. Urban operations are CSS-intensive, demanding large quantities of material and support for military forces and noncombatants displaced by operations.*
>
> FM 3-0

9-10. The LPT resembles and runs parallel to the intelligence preparation of the battlefield (IPB). Products generated under IPB may be useful in the logistic analysis. Conversely, the LPT may contribute to the IPB by identifying critical resources and infrastructure and assessing their potential to influence (positively or negatively) the operation plan. This information may warrant a course of action that includes offensive or defensive operations to seize, secure, or destroy those critical resources. In UO initially planned for other than logistic reasons, the information may require altering the plan or imposing additional constraints to protect the identified resources. These resources may or may not be critical to current operations; they are usually important to set or *shape* the conditions necessary for Army forces to *transition* to subsequent missions or redeploy. This close relationship between IPB and LPT underscores the need to quickly and continuously involve CSS personnel for their logistic expertise and perspective in planning UO.

URBAN LOGISTIC INFORMATION

9-11. Figure 9-2 illustrates that a thorough analysis of the key components of urban areas in the commander's AO provides the data for an accurate LPT and subsequent UO logistic support plan (see Chapter 2 and Appendix B). Analyzing the urban terrain and infrastructure helps to determine—

services purchased locally as fairly as possible among urban factions and ethnic groups to maintain impartiality and legitimacy. As part of their coordination efforts, commanders attempt to achieve the cooperation of relief agencies and other NGOs in this endeavor.

9-15. Finally, CSS planners also identify potential threats and increased protection requirements that the urban society (criminals, gangs, and riotous mobs) may present, particularly when CSS units and activities are located in urban areas. The disposition or allegiance of the urban population is also important to consider. The infrastructure of an urban area may exhibit great potential to support the logistic efforts of Army forces, but if the population is hostile or unreliable, the resources may be unavailable.

SUPPORT AREAS

9-16. A major influence on the operation plan and its subsequent execution is often the proper identification and preparation of support areas. The LPT helps commanders determine the need, advantages, and disadvantages of using urban areas in the AO as areas from which to provide support and conduct distribution operations. Ideally, these areas support reception, staging, onward movement, and integration operations. They allow easy sea and air access, offer adequate protection and storage space, facilitate the transfer of supplies and equipment, and are accessible to multiple LOCs. Consequently, commanders often establish support areas near seaports and airports that are part of a larger urban area. However, threats recognize the Army's need for ports and airfields and may devote substantial resources and combat power to defend them. Therefore, planners may determine during the LPT that the risks of seizing or establishing urban lodgment areas may be too high (see Chapter 4). Instead, they may recommend building an airfield, conducting logistics over-the-shore operations, or constructing logistic bases in more isolated locations.

OVERALL ASSESSMENT

9-17. As shown above, the LPT process and analysis help to determine if urban areas in the AO—

several others. Like urban infrastructure, they have two components: a physical component (supplies, equipment, and facilities) and a human component (the personnel who execute these functions). Like city mayors, commanders plan, manage, and synchronize these functions to provide responsive and efficient CSS for UO.

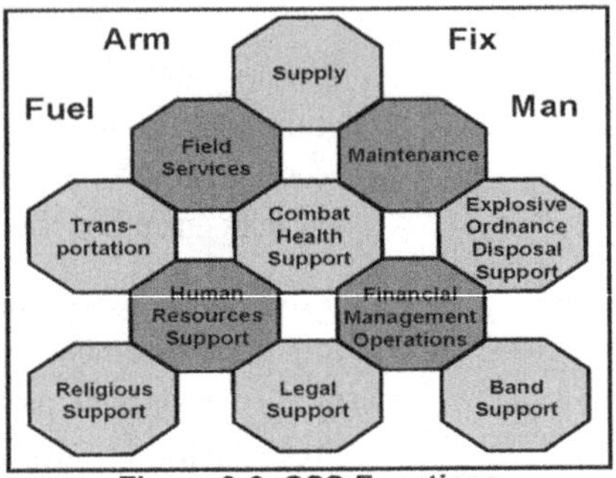

Figure 9-3. CSS Functions

9-19. Commanders and planners consider two essential aspects when addressing these CSS functions. One aspect looks outward and one looks inward. The first aspect is how these functions can best support full-spectrum UO—the outward analysis. The second aspect is how the urban area affects the conduct of CSS functions, particularly when those functions are located or performed in an urban area—the inward analysis.

SUPPLY

9-20. The supply function involves acquiring, managing, receiving, storing, protecting, maintaining, salvaging, and distributing all classes of supply (except Class VIII) required to equip and sustain Army (and joint) forces. In UO, commanders of major operations may need to make decisions early in the planning cycle. They decide whether to stockpile supplies forward or to rely on velocity management and scheduled and time-definite delivery to satisfy requirements. Some specialized items identified below may not be available through the normal military supply system and may take logisticians much longer to obtain or fabricate. Operation planners quickly identify the special equipment and increased supply requirements for UO to give logisticians time to acquire them.

> *Greater friendly force density would appear to make the providers' task easier. Logic would seem to dictate that more supported units in less space would translate to fewer nodes that require support, or at least less distance between a similar number of nodes than would be found on more open terrain. But the service supporter frequently finds the opposite is the case. . . . one is often not directly accessible from another due to enemy fires or physical barriers.*
>
> *On the Shoulders of Atlas*

Increased Urban Supply Requirements

9-21. Urban offensive and defensive operations will often increase requirements for Class V. Ammunition consumption rates have been five to ten times greater than operations in other environments. As such, urban combat operations will require a constant flow of ammunition. These operations will require more small arms and heavy machine guns; tank, antitank, mortar, attack helicopter, and field artillery ammunition (especially precision munitions); and mines, grenades, and demolitions.

9-22. With the potential exception of aviation fuel, Class III requirements for UO generally decrease at the maneuver unit level. Increased fuel requirements for engineer and power-generating equipment attached to or operating with forward units may offset these decreases. However, units may obtain tested fuel in the urban area (refinement facilities, gas stations, garages, and airfields).

9-23. Class IV will be in great demand in all types of UO to include large quantities of lumber (and the power tools to cut it) to reinforce positions and to bar access to windows and doorways. This lumber may come from outside the urban area or obtained by dismantling existing buildings and structures. Army forces use discretion, minimizing collateral damage, preserving critical infrastructure, and understanding the human dimension before they disassemble buildings. Some structures may be required to support operations or the civilian population during subsequent stages. Forces avoid structures of religious or cultural significance to prevent turning portions of the urban population against US operations and erode legitimacy. Other construction material will also be in high demand during relief operations to repair buildings and infrastructure damaged during the disaster. Class IV may also include specialized, prefabricated road barriers or collapsible wire-mesh or fabric cribs that can be filled with sand, rock, or dirt. They create barricades used to block roads, reinforce defensive positions, and protect headquarters and logistic activities.

9-24. Across the spectrum of UO, potable water may be a critical concern. Planners may need to ensure that additional containers for water (and fuel) are available to support dispersed stock at small-unit level. Units may need to increase levels of Class I and X supplies to support the urban population, particularly in urban stability operations and support operations. Soldiers may need more Class II, especially clothing and individual equipment, as exposure to the urban environment (concrete, glass, and steel) causes rapid wear. The chemical threats posed in UO may require pre-positioning large stocks of chemical protective clothing, defensive equipment, and decontamination apparatus. Units will also require items such as rope, grappling hooks, crowbars, ladders, chain saws, elbow- and kneepads, special vehicle and personnel body armor, fire-fighting equipment, packboards, and other specialized items to conduct operations.

Storage and Distribution

9-25. UO that are primarily stability operations or support operations may be able to rely entirely upon the Army's responsive, distribution-based CSS. Urban offensive and defensive operations can be part of an overall urban operation considered primarily a stability operation or support operation. In the past, these operations required both the attacker and the defender to prestock as many supplies as possible, particularly critical items. Defending forces recognize that the attacker will seek to isolate them from sources of supply; they ensure that they have enough supplies to execute the defense and restore their LOCs. Successful isolation of an urban area by the attacker ultimately results in a supply shortage and defeat. Complete isolation of a defender is difficult, especially if the urban area is large and the urban civilians support the defending forces. Nonetheless, attacking forces will seek, at a minimum, to interdict the defender's resupply operations. Although the isolation may be effective and the defense eventually fails, stockpiling supplies in this situation contributes to the defense.

9-26. Attacking forces may also want to stockpile supplies. This assists in maintaining the necessary tempo for a successful urban offensive. It can reduce the frequency with which supplies are moved over exposed LOCs. Stockpiling for offensive operations is not as common a technique as in the defense. Commanders review if they have the storage facilities and transportation assets (equipment and personnel) available.

9-27. Commanders weigh the benefits of stockpiling resources forward against relying on the Army's distribution-based sustainment system. Stockpiling brings supplies close to urban forces and helps ensure available supplies to support the tempo of UO. However, this method may burden the support structure that moves, handles, and protects large quantities of resources often on a repetitive basis. The Army's normal distribution-based CSS reduces this burden significantly; however, available transportation assets (sea, air, and ground) impacts delivery response times. These transportation assets often combine military assets overlaid on the host nation's (and urban area's) transportation and distribution infrastructure. Although the civilian infrastructure may initially support the Army's distribution system, later effects of UO, such as destruction of equipment and facilities or loss of civilian workers, may degrade the system. Army forces may have to share these assets with other military, civilian, and multinational forces or organizations participating in the urban operation, as well as with civilians. This shared system also puts at risk the timely delivery of critical supplies to Army forces.

9-28. CSS planners understand the urban environment and its effects on the proposed method of distribution. They also understand how urban operations (to include CSS UO) may affect the urban environment. Storing bulk fuel in or near an urban area, for example, may increase the risk of fire hazard to civilians and Army forces. With this awareness, planners present the force commander with an estimate that considers both risks and benefits. Depending on the particular area and other METT-TC factors, they may recommend one method of distribution later transitioning to another or a combination of methods. To be viable, the overall concept of logistic support, to include supply distribution, enables urban commanders to generate enough combat power to conduct decisive and shaping UO when and where required, and at a tempo faster than the threat can react.

9-29. During urban combat operations, the "push system" of supply distribution often works best to maintain the tempo of UO. Under this system, planners estimate supply requirements and arrange to have supplies delivered in preset packages (normally strategic, mission, unit, or combat configured loads). This method prevents critical delays of a "pull system" that requires units to request supplies and then await their arrival. The "best" method for tactical UO will be a combination of the two.

FIELD SERVICES

9-30. Field services involve feeding, clothing, and providing personal services for soldiers. It consists of food services, mortuary affairs, aerial delivery, laundry, shower, and textile repair. The urban commander determines the need and priority of each service after careful METT-TC analysis. Some facilities such as shower, laundry, and cold storage may be available in the urban area. Additionally, requirements to care for the urban population will increase requirements for field services immensely. In some circumstances, most notably urban support operations, field service units or activities will be critical and may be the only support provided.

Food Preparation

9-31. The feeding standard for soldiers operating in urban areas remains the same: three quality meals per day. Urban combat makes higher energy demands on soldiers who require a caloric intake of about 5,000 calories per day. Producing and delivering prepared meals to forward elements may be impossible or may be improved due to the urban environment. The area may contain facilities that aid in food storage and preparation that Army forces can use speeding transition to prepared rations. On the other hand, the threat situation (which may include the urban populace), rubble and other obstacles, and isolated and dispersed forces may prevent transporting and delivering prepared meals even if the capability to prepare meals is enhanced. Food (and water) may be available in the urban area; however, local sources must be tested, carefully monitored, and medically approved before consumption. Garbage disposal may be an important consideration in the urban area. Improper trash disposal may leave a signature trail (particularly during urban defensive operations) that may produce or worsen unsanitary conditions leading to increased disease and nonbattle injuries (DNBIs). Commanders understand that food operations, if not properly positioned and secured, can become a focal point for the urban population. Strict policy regarding distribution and control of any Class I supplies (including waste products) will be enforced. Black marketeers will be attracted to Army food service activities as well (even during offensive and defensive operations).

Water Purification

9-32. Water is essential; it is necessary for life, sanitation, food preparation, construction, and decontamination. Furnishing potable water is both a supply function and a field service. Water purification is a field service, normally performed with the storage and distribution of potable water—a supply function.

9-33. **Vigilant Monitoring.** Urban areas will often have a ready source of water to support the urban inhabitants and its infrastructure. However, this water may not be potable for US and allied forces (though the urban population may have developed immunity toward its microorganisms). The higher concentration of TIM in urban areas compounds this problem, as ground water is highly susceptible to chemical contamination, even supplies located miles away from the source of contamination. US soldiers are trained and cautioned against using water from an urban area (to include ice and bottled water) until preventive medicine and veterinary personnel can determine its quality. When water quality is unknown, commanders ensure use of tactical water purification equipment that will unquestionably upgrade it to Army water quality standards. Even if initial testing indicates the urban water is safe for Army forces, personnel continuously monitor the water quality. However, Army water purification, storage, and transportation requirements for UO can be greatly reduced if the existing urban water supply can be integrated into CSS operations. An early assessment of the feasibility of this course of action is critical to CSS planning.

9-34. **Greater Requirements.** Individual water requirements are greater for soldiers operating in an urban environment due to increased levels of exertion. Offensive and defensive UO are often intense and can produce more casualties, including civilians. Consequently, medical facilities, already consumers of large volumes of water, may require even more water. Water purification, particularly in the urban areas of developing nations and during urban disaster relief operations, will be a critical and constant concern for Army forces.

9-35. **Potential Key Terrain.** Sites that can control the water of the urban area may be key terrain, providing not only a resource for Army forces, but also a means to control the threat, the civilians, or both. These sites may be the sources of the water—the river, lake, reservoir, or storage tanks—or the means that process and transport the water—pipelines, pumping stations, or treatment facilities. Many sites may be outside the urban area, as many large urban areas draw water from distant sources. The seizure of a pumping station or pipeline may make it possible for commanders to control water supplies without expending resources required to enter the urban area. To preserve critical infrastructure, commanders may increase security to protect these locations from contamination or destruction. Engineers also may need to restore, maintain, or operate existing water facilities damaged by the threat or disaster and to drill new wells and construct new water facilities.

Mortuary Affairs

9-36. Mortuary affairs provide the necessary care and disposition of deceased personnel. It supports the Army across the spectrum of operations. It may directly and suddenly impact (positively or negatively) the morale of the soldiers and the American public and may influence relations with the civilian population in the AO. It can also affect the health of soldiers and the urban populace. Commanders plan evacuation routes and temporary collection and internment sites, trying to adhere to local customs and traditions to lessen potential negative consequences.

9-37. Units are responsible for recovering the remains of their own fatalities and evacuating them to the closest mortuary affairs collection point, usually located at the nearest support area. Because of the density of noncombatants in UO, commanders may also find themselves responsible for civilian remains. Deaths of civilians under Army control, such as urban evacuees and refugees at Army-operated sites, often obligate the Army to care for their remains including medical certification and records of death. High-intensity urban combat may result in civilian deaths, and health concerns will require Army forces to deal with civilian remains expeditiously. Commanders consult local religious leaders, the Staff Judge Advocate (SJA), CA personnel, and chaplains to verify that they are abiding by law and customs. Overall, commanders ensure that forces treat all deceased, including civilians, with dignity and respect (another important aspect of adhering to the urban fundamental of understanding the human dimension).

Aerial Delivery

9-38. Aerial delivery is the movement by fixed- or rotary-wing aircraft and delivery by the use of parachute or sling load of soldiers, supplies, and equipment. As a vital link in the distribution system, it adds flexibility and provides the capability of supplying the force even when urban ground LOCs are disrupted. Forces use aerial delivery to deliver supplies and equipment when no other means can. However, in support operations it is used extensively to move supplies to meet the urgent needs of a population in crisis. In all UO, delivery aircraft are highly vulnerable to small arms, rockets, and air defense systems. A threat may further decrease an already limited number of urban drop zones (aircraft may be able to avoid air defense systems, but ground forces may not be able to secure the drop zone and retrieve the cargo). Equipment and supplies transported by helicopter sling-load lessen the latter disadvantage. There are usually more available sites to deposit sling loads, such as rooftops (engineers will be critical in determining the structural integrity of rooftops for landing helicopters), parking lots, and athletic fields. However, load instability during flight may restrict a helicopter's airspeed and maneuver capabilities making it more vulnerable to small arms and man-portable air defense systems. For all these reasons, aerial delivery of supplies in UO may be much less efficient than in many other environments.

Laundry, Shower, and Light Textile Repair

9-39. Soldiers are provided clean, serviceable clothing and showers for hygiene and morale. A field services company provides direct support at the tactical level. During UO, other sources such as fixed urban facilities obtained via host-nation support and contract services may provide these services.

MAINTENANCE

9-40. Maintenance entails keeping materiel in operating condition, returning it to service, or updating and upgrading its capability. It includes recovering and evacuating disabled equipment; replacing forward; performing preventive maintenance checks and services (PMCS); increasing battle damage assessment and repair (BDAR); and analyze potential resources for maintenance in the urban area.

Rapid Recovery Essential

9-41. Disabled vehicles easily block narrow thoroughfares during urban offensive, defensive, and some stability operations. This makes rapid recovery operations essential. Hastily secured unit maintenance collection sites near the damaged equipment and along supply routes are necessary to avoid clogging limited LOCs and mounted avenues with vehicle evacuation operations. The task organization of armored units into smaller attachments (often platoon-sized or smaller) will strain limited recovery assets. Units maintain centralized and responsive control over these potentially critical resources.

Replace Forward

9-42. One of the guiding maintenance principles is to replace forward and fix rear. Maintenance activities, with a forward focus on system replacement, task and use the distribution and evacuation channels to push components and end items to the sustainment level for repair. However, the conditions of UO may make distribution and evacuation difficult. Fixing equipment on site is extremely important in UO. Organizational maintenance personnel accurately evaluate damage to their equipment. Recovery of equipment will prove difficult. When recovery is required, equipment is moved only as far rearward as the point where repairs can be made. When selecting a maintenance site, commanders consider: security, a sufficient area around equipment for lift or recovery vehicles, and use of existing maintenance facilities or garages.

9-43. The unforgiving urban terrain will invariably increase damage to man-portable weapons and equipment, particularly electronic equipment sensitive to jarring. Although by definition man-portable weapons and equipment are easier to transport, evacuating these systems may prove as difficult as evacuating vehicles and larger, heavier equipment. Therefore, unit or direct support maintenance support teams (MSTs) will frequently need to repair equipment at (or as near as possible to) the point where it was damaged. Equipment operators are responsible to properly diagnose the fault or damage. Such action ensures that the correct repair parts and maintenance personnel are sent forward to complete necessary repairs. In UO, particularly offensive and defensive operations, units may need to replace rather than repair equipment, requiring CSS personnel to plan for increased replacement of what might normally be repairable equipment, as well as increased repair parts for man-portable items.

Add-On Protection Increases Wear

9-44. Maintenance units may need to attach additional armor to both wheeled and tracked vehicles operating in an urban environment to increase protection against small arms, mines, rocket-propelled grenades, and lightweight antiarmor weapons. (Units can also sandbag vehicles to achieve a degree of increased protection.) These modified vehicles, however, may put excessive wear on brakes, springs, suspension, and tires (already vulnerable to the increased amount of debris caused by many UO). After several months, these same vehicles may experience severe damage to major assemblies, such as engines and transmissions. The increased protection proportionally increases repair parts and command emphasis toward inspecting these items during daily operator PMCS. Commanders employing this additional protection may also consider increasing the number of scheduled services.

Increased Battle Damage Assessment and Repair

9-45. In UO, operators, crews, MSTs, and recovery teams execute BDAR far more than in other environments. BDAR quickly restores minimum essential combat capabilities for a specific mission (normally of short duration) or allows the equipment to self-recover by expediently fixing or bypassing components. Commanders may need to authorize supervised battlefield cannibalization and controlled exchange when units lack critical parts or cannot bring them forward.

Potential Urban Maintenance Resources

9-46. Although urban areas can complicate maintenance, they may contribute to this CSS function. Analyzing the urban area in the commander's AO may reveal potential sources of parts, tools, equipment, and facilities necessary to fix equipment and fabricate critical parts. Urban areas may serve as key sources for parts and facilities (and contract personnel) to repair automation and network communication equipment.

TRANSPORTATION

9-47. Transportation supports the concept of the urban operation by moving and transferring units, soldiers, equipment, and supplies. Transportation incorporates military, commercial, supporting nation, and urban area capabilities to build a system that expands to meet the needs of the force. Transportation includes movement control, terminal operations, and mode operations.

9-48. Urban areas are often critical to transportation operations. These areas may serve as a lodgment or support area for entry of Army forces and sustainment supplies. The existing transportation and distribution infrastructure may be essential to reception, staging, and onward movement. Contracts and host-nation support agreements may greatly increase the ability of Army forces to use the urban area's facilities, which may include docks, airfields, warehouses, and cargo handling equipment. Urban support may also include skilled urban workers, such as longshoremen and MHE operators.

Urban Terminals

9-49. In addition to serving as major seaports and aerial ports of debarkation, other urban areas may provide additional terminals in the AO. Forces may use these terminals for further staging, loading, discharging, and transferring the handling between various inland transportation modes and carriers (motor, air, rail, and water). These urban terminals—with synchronized movement management—permit commanders to rapidly shift transportation modes and carriers. Such action increases flexibility and ensures the continued forward movement of equipment and supplies to influence the tactical situation. Movement control, particularly in urban areas, relies heavily on support from military police in their maneuver and mobility support role. Without this support, urban LOCs may become congested, hinder movement and maneuver, and degrade force effectiveness (see FM 3-19.4). Urban commanders may need to establish multiple roadblocks and traffic control posts, restrict selected roads to military traffic, and reroute movement to unaffected road networks when civil support and refugee control operations compete for available routes. Military police operations are critical in this regard and will require continuous, close coordination with urban civilian police.

Obstacles to Ground Transportation

9-50. Although urban areas can contribute to transportation operations, rubble and other damage can become obstacles to ground movement. Even in an undamaged urban environment, road and bridge weight restrictions may limit transportation operations. Urban route maintenance, to include reinforcing bridges, may become a priority task for engineer units. Bypassed pockets of resistance and ambushes pose a constant threat along urban supply routes. Urban LOCs will often require increased security in the form of continuous route security operations, regular (daily, if necessary) mine clearance operations, numerous observation posts, and a larger, more mobile tactical combat force. Such security increases manpower requirements for sustaining operations and potentially reduces resources from decisive operations. Moving critical supplies may require heavily armed convoys or lightly armored vehicles instead of trucks. Drivers are well trained, rehearsed, and alert. They can recognize and avoid potential mines and minefields (such as driving in the same tracks as the vehicle in front) and can react rapidly to ambushes. (In addition to the measures above, Russian convoys during their operations in Chechnya were not allowed to move without attack helicopter escort and the availability of immediate close air support.) Aerial resupply alleviates problems due to ground obstacles, but the air defense threat and proximity of threat forces may preclude their routine use.

Population Effects

9-51. The ability of Army forces to use vital urban transportation facilities depends largely on the civilians and the threat. The civilian population can affect the transportation system if they do not support the goals of Army operations. Urban transportation systems—such as ports, railroads, and rivers—require many specialists to operate. Without these specialists, the system's utility is degraded and may not function at all. In urban stability operations or support operations, Army forces will share the system with civilians and other agencies. Civilian authorities may refuse to allow Army forces to use any portions of an urban area's transportation system. Negotiating for access to that system under these circumstances then becomes a command priority.

Threat Effects

9-52. The threat can significantly affect urban transportation systems. Many are composed of smaller subsystems. Each subsystem is vulnerable to attack, which in turn often shuts down the whole system. A large canal system, for example, may have entrance and exit facilities, the canal itself, a means to pull the vessel along such as a locomotive engine, and the civilians that run each of these subsystems. Both an attacker and defender understand the components of the particular transportation system. If important to their current or subsequent operations, defending forces then develop plans and allocate forces to protect these subsystems. Attacking forces, on the other hand, often avoid collateral damage to the system, while simultaneously preventing enemy destruction of the facilities.

COMBAT HEALTH SUPPORT

9-53. Combat health support (CHS)—

Care of Civilians

9-55. In combat operations, the military normally does not provide injured civilians with medical care. Saving civilians is the responsibility of civilian authorities rather than the military. However, based on METT-TC and requirements under Geneva Conventions, commanders may need to recover, evacuate, and treat numerous civilians (particularly in urban support operations and some stability operations) until the local civilian medical personnel and facilities can be reconstituted and supplied. In urban support operations involving weapons of mass destruction, the primary focus of Army support may be CHS. If commanders provide medical support, they adhere to the UO fundamental of transitioning control and transferring responsibilities for medical care of civilians to another agency or into civilian hands as soon as is practical. In UO, commanders consider and address the medical treatment of civilians (enemy and friendly) early in the planning process. Any initial assessment or survey teams therefore contain CHS representatives. This assessment also considers cultural factors related to civilian medical treatment. Commanders may need to develop specific medical policies, directives, and standing operating procedures to ensure that subordinates know how much medical care they may provide to—

prevent introducing new diseases (especially drug-resistant strains) into an urban area. A new disease may tax the medical system and introduce a new medical problem into an area already in crisis. Lastly, field sanitation training (to include training in the use of barrier protection such as latex gloves when rendering care to any person and animal, rodent, and pest control), equipment, and supplies are part of overall preventive medicine measures and considerations.

Combat Stress

9-58. Stress occurs in every operation and type of environment; some stress is beneficial but too much is harmful. Controlled combat stress can invoke positive, adaptive reactions such as loyalty, selflessness, and heroism. On the other hand, uncontrolled combat stress can result in negative, harmful behavior and combat stress casualties. Such behaviors and casualties—battle fatigue, misconduct stress behaviors, and post-traumatic stress disorder—can interfere with the unit's mission. Physical and mental factors leading to combat stress result from the environment and the deliberate enemy actions aimed at killing, wounding, or demoralizing soldiers. However, many stressors are also generated from the soldier's own leaders and mission demands.

Combat Stress
Chechnya – 1994 to 1996

Russia's 1994-1996 conflict with Chechnya, a republic in the southwestern part of the Russian Federation, produced an increased number of psychological trauma and combat stress casualties. One medical survey found 72 percent of the soldiers screened had some sort of psychological disorder symptoms. Of that, 46 percent exhibited asthenic depression (a weak, apathetic, or retarded motor state). The other 26 percent exhibited psychotic reactions such as high states of anxiety, excitement, or aggressiveness and a deterioration of moral values or interpersonal relations. The statistics showed more troops experienced combat stress disorders than during their 1980s war in Afghanistan. One primary difference was that in Chechnya, Russian forces conducted combat mostly in cities rather than in mountains, valleys, and other rural areas.

Combat always invokes fear in soldiers. However, poor training and planning, uncertainty in their cause, and urban populations that resented their presence exacerbated the psychological climate for the Russian forces in Chechnya. Acts of subversion and terrorism by Chechen guerrillas kept the Russians in a constant, high state of readiness and anxiety; the soldiers viewed every civilian—young or old, male or female—as a potential enemy. This psychological pressure was not simply a by-product, but an objective of information operations and a prime reason for taking the fight into the close confines of Chechnya's urban areas. Torture and mutilation of prisoners; immediate execution of captured pilots; imitative electromagnetic deception (Chechens mimicked Russian radio transmissions and directed Russian close air support against their own forces); and propaganda to convince civilians that Russia's actions had a religious bias against Muslims and Islam were conducted to exert intense, unremitting psychological pressure on Russian forces—with great success.

The characteristics of urban areas combined with Chechen insurgent activities and information operations, civilians that did not welcome foreigners, and an unpopular and poorly supported conflict with an open-ended mission reinforced the need for the national will to strengthen and support the fight when forces deploy to urban areas. These characteristics also reinforce the need for clear objectives, proper leadership and training, and available medical assets able to properly diagnose and treat combat stress casualties.

9-59. In offensive and defensive UO, compartmented urban combat leads to physical isolation, difficulties in transmitting radio signals lead to communications isolation, and combined they create an overwhelming sense of being alone. Snipers, mines, and booby traps combined with the closeness and high intensity of urban combat contribute to an unremitting fear of attack from any quarter that further increases stress casualties. Additionally, seeing and perhaps accidentally inflicting casualties on civilians (especially women and children) increases battle fatigue. If civilians are hostile or a threat uses the population as cover and concealment, then the potential for misconduct stress behaviors often increases. Urban areas may provide temptations for looting, alcohol and substance abuse, black marketeering, and harmful social interactions; these temptations may increase misconduct stress behaviors. Leaders can prevent or rapidly identify, successfully manage, and treat stress-related casualties (see FM 22-51 and FM 6-22.5) and prevent misconduct stress behaviors as well as potential violations of the law of war. They provide training (both in urban combat and combat stress management), effective rules of engagement (ROE), unit cohesion, strong leadership, and mental health support. Leaders, forward on the battlefield, will determine whether discipline perseveres.

Evacuation

9-60. Transportation restrictions may preclude evacuation of urban casualties. Consequently, units may require more litter bearers to move the injured to a point where they can be further evacuated by ground or air ambulance. Lengthy evacuation routes will require more litter bearers, as multiple litter relay teams will be necessary to conserve energy and expedite evacuation. However, unless augmented or relieved of this responsibility by another unit, these litter bearers will come from the casualties' own unit thereby diminishing the unit's strength levels necessary to accomplish its primary mission. Depending on the expected level of casualties, commanders may augment units with additional personnel to perform evacuation or may assign maneuver units this mission. Historically, urban offensive and defensive operations experienced more casualties than these operations in other environments. Evacuation personnel will require specific training in urban evacuation techniques (moving casualties from subsurface and supersurface levels to and along the surface level). Overall, urban commanders at all echelons develop detailed medical evacuation plans. Engineers are critical to clear routes for medical resupply and evacuation. Army forces may need the unique capabilities, equipment, and skills now typically seen in civilian urban search-and-rescue teams to clear debris and search for casualties. Specially trained dogs may also play a vital role in locating victims.

Treatment

9-61. First aid training will have increased significance in UO. The compartmented nature of UO, transportation restrictions, communications difficulties, and the finite number of combat medics may limit the urban casualty's initial treatment administered by nonmedical personnel or to self-treatment measures. Units identify and train combat lifesavers to perform in the absence of medics. Since the likelihood of Army forces performing UO continues to increase, commanders strive to meet or exceed Army standards for the number of combat lifesavers required for their specific unit. This increase in self, buddy, and combat lifesaver care, as well as longer delays in evacuation, may also increase requirements for additional first aid and medic-carried supplies.

9-62. In addition, the increased potential for delayed evacuation during UO mandates that Army combat medics be skilled in prolonged casualty care. (During the 3-4 October 1993 battle in Somalia, seven medics managed 39 casualties for more than 14 hours before they could be evacuated.) Evacuation delays significantly increase potential infection. Such delays may cause more casualties dying of their wounds; therefore, combat medics should also be skilled in administering antibiotics on the battlefield.

9-63. All CHS personnel can recognize and treat injuries due to incendiary or fuel-air explosives (also known as thermobaric weapons)—a favored, urban-oriented threat weapon (see Chapter 3). These weapons explode; create a cloud of volatile gases, liquids, or powders; and then ignite, creating an immense fireball consuming oxygen and creating enormous overpressure. When employed in an urban structure, the blast wave or overpressure is greatly amplified. Injuries resulting from these weapons are massive burns, broken or crushed bones, concussions, missile injuries, and internal injuries. Medics or doctors can easily overlook internal injuries (at least initially) unless they are trained, prepared, and expecting them.

9-64. The increased use of body armor during UO will help prevent penetrating chest and abdomen wounds. CHS personnel expect more groin, pelvis, and extremity injuries. Furthermore, when fighting soldiers that are known to use body armor, a threat (particularly snipers) can be expected to target the head and face more often than other anatomic areas resulting in more head injuries. Lastly, hearing loss may increase particularly when firing recoilless weapons in enclosed spaces with little ventilation.

EXPLOSIVE ORDNANCE DISPOSAL SUPPORT

9-65. Explosive ordnance disposal (EOD) support provides the capability to neutralize domestic or foreign conventional, nuclear, chemical, and biological munitions and improvised devices. Such devices threaten military operations and military and civilian facilities, materiel, and personnel. Unexploded explosive ordnance (UXO) creates a much greater risk during UO than operations in any other environment. Confined spaces, hard surfaces, and more personnel (both soldiers and civilians) in the vicinity may magnify the detonating effects of UXO. Dense terrain makes UXO more difficult to locate. In fact, terrorists select urban settings for their improvised explosive devices (IEDs) to potentially kill and destroy more, thus gaining greater visibility for their message. EOD units perform many tasks (detecting, identifying, rendering safe from, and disposing of explosives) associated with UXO and IEDs. Urban operations will rely more heavily on their role as advisors and instructors on UXO hazards, protection measures, and disposal techniques. EOD specialists will advise and train other Army forces, other services, multinational partners, and civilian authorities. EOD specialists will often work closely with public affairs and psychological operations personnel to increase awareness and teach the urban population to identify and avoid UXO.

HUMAN RESOURCES SUPPORT

9-66. Human resources support (HRS) encompasses the following functions: manning the force, personnel support, and personnel services. These activities include personnel accounting; casualty management; essential services; postal operations; and morale, welfare, and recreation provided to soldiers, their families, Department of Army civilians, and contractors.

9-67. Successful UO require HRS functions. Clearly, HRS focuses on caring for the needs of people: soldiers and the civilians who support them. Since a critical component of the urban environment is the population, these activities, when required, may also support them. For example, personnel elements may support accountability of displaced persons and civilian internees. In conjunction with mortuary affairs, casualty managers may assist with civilian death records and reports (always maintaining sensitivity to the confidentiality of casualty information). A postal company may assist urban officials in training and reestablishing civilian postal operations. Personnel elements may also provide identification cards and documents to support increased populace and resources control measures.

> *Urban operations will continue to be manpower intensive, even as advances in technology make the conduct of urban battle more precise and discriminating.*
>
> FM 1

9-68. In offensive and defensive UO, HRS personnel will need to account for more casualties and more frequent reconstitution. In all UO, success relies on training individual replacements. Urban combat requires soldiers skilled in specific urban combat tactics, techniques, and procedures. They understand the societal aspects of the urban population and have training in crowd control to avoid escalating potentially explosive situations. Training these replacements while in the replacement system frees urban commanders from having to do so. It also helps reduce soldier (and supporting civilian) isolation, anxiety, and fear associated with urban areas and operations.

FINANCIAL MANAGEMENT OPERATIONS

9-69. Financial management operations make resources available when and where they are needed and assist the urban commander in maintaining fiscal responsibilities. Finance operations are necessary to conduct contracting and provide real-time information, accounting, and services. Resource management operations ensure that urban operational policies and procedures adhere to laws and regulations, develop command resource requirements, and leverage appropriate fund sources to meet them.

9-70. UO will likely include other US military services, governmental agencies, and contractors; multinational forces; NGOs; and various elements of the host nation. So many actors complicate financial accounting, resource management, and cost capturing. Yet, these activities are vital to accomplishing UO and maintaining legal requirements. Army forces will receive support from and provide support to these participants. Resource managers provide crucial "in-theater" expertise to the urban commander on the obligation authority. They also furnish the documentation necessary to obtain and pay for local goods and services using contract or commercial vendor services. However, commanders (with finance unit assistance) assess the economic impact of UO on the economy. Analysis includes how well the urban economics and commerce infrastructure can support the deployed force and how an influx of US currency may affect the overall economy.

9-71. Finance units can also provide joint pay support and non-US pay support for host-nation employees and day laborers supporting UO, as well as for civilian internees. These units can support bounty programs and solatia (financial compensation). Commanders use bounty programs to purchase weapons, radios, information, and other items from the urban inhabitants. They also use caution when paying for critical human intelligence. First, they ensure that multiple collection agencies in the command avoid paying the same source for the same information and interpreting these repeated inputs as validation that information is accurate and reliable. Second, they establish a price or pay scale so subordinate units (particularly in multinational UO) equally compensate their sources for information. Commanders may also need to make solatia to alleviate grief, suffering, and anxiety resulting from injuries and property or personal loss resulting from some aspect of UO. At other times, commanders make solatia to meet cultural expectations of the urban population. These payments are nominal in amount and without an admission of liability by the Army.

RELIGIOUS SUPPORT

9-72. Religious support entails providing and performing religious support operations for commanders to protect the free exercise of religion for soldiers, family members, and authorized civilians. This includes personal delivery of rites, sacraments, ordinances, pastoral and spiritual care, and religious education. Such support also consists of advising the commander on matters of religion, morals, and morale as affected by religion; the impact of local religions on the mission; and the ethical impact of command decisions.

Moral and Ethical Climate

9-73. Chaplains can help commanders sense the moral and ethical climate in Army units and understand potential moral dilemmas associated with planned UO. The urban environment affects soldiers' mental health and can increase combat stress casualties, especially misconduct stress behaviors. Chaplains are an important part of combat stress support. The chaplain's presence and faith sustains soldiers throughout periods of great trauma. Religious support contributes to the total well being of soldiers and aids in their return to combat readiness. Chaplains help bolster soldiers' moral and ethical behaviors through spiritual fitness training so that soldiers may better cope with ambiguous moral and ethical situations. Chaplains can help identify ethical concerns before they become critical command problems. To this end, their observations help develop and modify ROE, which—due to ambiguity or overly restraining rules—may be part of the problem.

Influence of Local Religions

9-74. Chaplains advise commanders on matters of religion as it affects the soldiers within their units. They also explain the influences of local religions on the urban populace and their potential effects on Army forces and UO. Religion is a crucial aspect of assessing the societal component of the urban environment. Understanding the major tenets and concepts of the religions and the impact of faith on civilians' lives may help commanders understand what motivates the populace. This understanding also helps commanders to appreciate the inhabitants' attitudes toward other races, religions, and cultures and to identify unacceptable kinds of social interaction (particularly between soldiers and civilians). Failure to recognize and respect religious beliefs can rapidly erode the legitimacy of the mission. A thorough analysis of the urban environment also includes the degree of influence religion and religious leaders have on the area's government, military, and economy.

LEGAL SUPPORT

9-75. Legal support provides operational law support in all legal disciplines (including military justice, international law, civil law—comprised of contract, fiscal, and environmental law, claims, and legal assistance). This support assists in command and control, sustainment, and HRS of UO. Legal considerations are important in any operation; they take on added significance during UO. They form the foundation for establishing ROE and are critical in the targeting process (determining protected targets, for example). They affect how units acquire goods and services from urban areas and provide support to other agencies and organizations operating in an area. The environment's complex nature requires commanders and their staffs to review and closely consider applicable legal constraints when developing and executing courses of action. Most urban areas have a highly developed legal system. The SJA support to commanders address this urban system and its potential to affect, positively or negatively, UO.

9-76. International, host-nation, and US law and other regulatory guidelines may vary in their applicability by time and place; actions permissible in one situation may be prohibited in another. These exceptions and complexities increase requirements for SJA, often working with CA personnel, to identify and resolve technical legal issues. Therefore, the SJA actively advises and participates in all aspects of UO from predeployment training and initial planning through transition and redeployment. FM 27-100 contains detailed legal guidelines affecting UO.

International and Host-Nation Law

9-77. International law consists primarily of agreements, treaties, and customary law to include the law known as *the law of war* (see FM 27-10). The law of war consists of four general principles applicable when conducting any operation but requiring particular attention during UO. Figure 9-4 lists the four principles: military necessity, discrimination (or distinction), unnecessary suffering (or humanity), and proportionality.

Military Necessity: The principle that justifies the use of measures not forbidden by international or domestic law necessary to rapidly achieve military objectives.

Discrimination: The principle of distinguishing between combatants (who may be attacked) and noncombatants (who may not be attacked).

Unnecessary Suffering: The principle that prohibits the use of weapons, projectiles, or other materials in a manner calculated to cause superfluous injury and unnecessary suffering.

Proportionality: The principle that injury to persons and damage to property incidental to military action, in the circumstances ruling at the time, must not be excessive in relation to the concrete and direct military advantage anticipated.

Figure 9-4. General Principles of the Law of War

9-78. International law may affect urban operational issues, such as the right of entry, base operations, use of urban infrastructure, and overflight and landing rights. Status-of-forces agreements (SOFAs) exist or can be negotiated to resolve legal issues, such as the status of soldiers operating in foreign areas to include criminal and civil jurisdiction, taxation concerns, and claims for damages and injuries. Unless a SOFA or other convention exists, soldiers operating in foreign urban areas have the same legal status as tourists; they are subject to the laws and judicial process of the host nation. Commanders are responsible to understand the international and host-nation agreements and laws that influence foreign UO. If local law hinders the operation, commanders may be able to inform the local US diplomatic mission and request that it negotiates a solution.

US Law

9-79. UO also comply with US law whether it is in the form of a statute, executive order, regulation, or other directive from a federal branch or agency. US law influences UO by governing the acquisition of supplies and services for Army forces, regulating the assistance that can be rendered to foreign nations, and controlling intelligence activities. The Posse Comitatus Act, for example, makes it a crime for Army forces to enforce civil law. Similarly, portions of the Foreign Assistance Act prohibit soldiers from performing law enforcement activities in foreign urban areas. However, circumstances—expressly authorized by the Constitution, acts of Congress, and other exceptions to these statutes—exist that allow the Army to support civilian law enforcement. Although not nearly all-inclusive, the above demonstrates how US law further complicates urban operations, particularly stability and support, and increases the need for SJA advice and counsel in all facets of UO.

Legal Aspects of Nonlethal Force

9-80. Nonlethal capabilities can augment the means of deadly force and extend urban firepower options. They enhance the commander's ability to apply force in proportion to the threat and to allow discrimination in its use. The range of nonlethal capabilities includes offensive information operations, smoke and obscurants, irritants (such as chemical riot control agents), non-penetrating projectiles, and high-pressure water devices. These continually expanding capabilities give commanders more options to confront situations that do not warrant deadly force but require soldiers to employ overwhelming decisive power. However, nonlethal capabilities are subject to the same legal constraints as lethal force (in fact, some nonlethal capabilities can cause serious injury and death, particularly if not employed properly) and undergo the same legal review. Like lethal force, nonlethal capabilities show military necessity, distinguish between combatants and noncombatants, distinguish between military objectives and protected property, are used proportionally, and do not result in unnecessary suffering. Commanders cannot employ chemical herbicides or riot control agents without prior presidential approval.

BAND SUPPORT

9-81. Bands provide music to instill in Army forces the will to fight and win. Bands also foster the support of multinational partners and urban populations through support to ceremonies, troop support functions, concerts, protocol functions, and religious ceremonies. Army bands quickly and effectively communicate professionalism, a positive image, and a nonthreatening show of force. Throughout the spectrum of operations, planners consider band support when developing their information operations concept of support to UO. Because UO are often resource intensive, band members may be required to assist in other essential activities unrelated to music. Such tasks may include casualty evacuation; command post security; and NBC reconnaissance and decontamination.

GENERAL ENGINEER SUPPORT

9-82. General engineer support will be essential during UO. This support helps assess, construct, maintain, and restore essential LOCs and urban facilities to sustain Army forces, the urban population, or both. Using civilian resources and investing Army general engineer resources requires careful consideration by commanders and staff planners. Since all elements of the urban infrastructure interconnect, general engineering support touches each category to some degree. Figure 9-5 illustrates how urban-specific, general engineering tasks align primarily with the transportation and distribution, energy, and administration and human services components of the urban infrastructure. These engineering tasks are significant and readily apply to UO. The last two, providing fire fighting support and waste management, have not been previously addressed and require more specific consideration.

THE USE OR INVESTMENT OF RESOURCES

Construct, Maintain, or Restore

Figure 9-5. General Engineer Support

9-83. During urban offensive and defensive operations, Army engineer units accomplish tasks to sustain or improve mobility, survivability, and sustainability of US and allied forces. These units maximize the existing urban facilities, host-nation support, civilian contractors, and joint engineer assets. Commanders consider how using urban facilities to support military forces may negatively affect the population. On the other hand, construction and repair may benefit both Army units and the urban inhabitants. Restoring the urban transportation network not only improves military LOCs, but may also allow needed commerce to resume. Repairing urban airfields or ports increases throughput capabilities for military supplies, facilitates medical evacuation operations to the support base, accelerates needed relief efforts, and allows international commerce to proceed. Commanders may first invest resources and conduct general engineering tasks to restore facilities for civilian use. Such actions stem future drains on operational resources or to facilitate later transition of control back to civilian authorities. For example, repairing police stations, detention facilities, and marksmanship ranges may help urban governments reestablish law and order after completing urban offensive or defensive operations. During most support operations and some stability operations, the focus of general engineering clearly supports and assists the urban population rather than sustains Army forces.

9-86. A military force task organized with multiple fire fighting teams (even with maximum use of available civilian fire-fighting assets) will only be able to fight some fires in the AO. Water distribution systems damaged during operations, chemical and other TIM, and hostile activities will further complicate and limit fire fighting capabilities. Commanders develop priorities for equipment, facility, and infrastructure protection. All soldiers need training in fire prevention and initial or immediate response fire fighting. Such training includes planning covered and concealed movement, withdrawal, and evacuation routes. Soldiers are trained to identify and remove ignition and fuel sources and be provided additional fire fighting material such as extinguishers, sand, and blankets (see FM 5-415).

WASTE MANAGEMENT

9-87. Management of all forms of waste, particularly human, putrescible (such as food), and medical, may become a critical planning consideration for Army forces. This particularly applies if the urban waste management infrastructure was previously inadequate or damaged during UO, the Army force is operating in the urban area for an extended period of time, and a significant number of the urban population remains. Failure to adequately consider this aspect, possibly coupled with an inadequate water supply, may create unacceptable sanitary and hygiene conditions and subsequently increase DNBIs as well as civilian casualties.

CIVIL-MILITARY OPERATIONS

9-88. Commanders use civil-military operations (CMO) to establish, maintain, influence, or exploit relations to achieve operational objectives. These relations are among military forces, civilian authorities and organizations (both governmental and nongovernmental), and the civilian population. Because of the urban society's link to other aspects of the urban environment, CMO will prevail in all UO. Commanders may need to assume temporary responsibility for functions and capacities of the urban government. Although unstated, civil-military cooperation, particularly during urban stability operations and support operations, will be an essential, implied task of the mission. Like public affairs, effective CMO is based on establishing and maintaining credibility and trust with the urban populace and civilian organizations operating in the urban environment.

CIVIL AFFAIRS

9-89. CA units will be critical during UO. While any military force can conduct CMO, CA units are specifically organized, trained, and equipped to conduct activities in support of CMO. They have experience in planning and conducting CMO, a regional focus (which includes enhanced cultural awareness and language training), and civilian technical expertise. Such experience ensures relevant support to commanders conducting urban operations. CA units organize their capabilities into 16 functional skills normally arranged in four specialty function teams (see Figure 9-6). Commanders use these skills, often unfamiliar to most military personnel, to—

Urban Combat Service Support

interact with the civilian community. Examples of information available to CA units include government documents, libraries, and archives; files of newspapers and periodicals; industrial and commercial records; and technical equipment, blueprints, and plans.

ASSESSMENT OF CIVIL CONSIDERATIONS

9-92. As part of the initial planning process, CA units conduct an area assessment, which can provide commanders with essential information about the environment. Commanders integrate this initial assessment into the overall urban IPB process (see Appendix C, FM 41-10). To help analyze civil considerations, commanders and staffs can consider many characteristics such as physical terrain, structures, capabilities, organizations, people, and events. These characteristics easily align with terrain, society, and infrastructure; and, like them, they are overlapped and interdependent (see Figure 9-7 and FM 6-0).

9-93. Overall, CA personnel help commanders understand the complexities of the infrastructure and societal components of the urban area. These components (together with the terrain or physical component of the urban area) interconnect. CA forces help identify and understand the relationships and interactions between these urban components. From this understanding, commanders can anticipate how specific military actions affect the urban environment and the subsequent reactions. CA personnel consider the short-term effects and reactions as well as the long-term consequences. Understanding these long-term consequences helps ensure a smooth transition of the urban area back to civilian control.

Appendix A

Siege of Beirut: An Illustration of the Fundamentals of Urban Operations

The IDF had neither the strategy nor the experience nor the configuration of forces to fight and sustain a house-to-house campaign in Beirut.

Richard A. Gabriel
Operation Peace for Galilee: The Israeli-PLO War in Lebanon

OVERALL STRATEGIC SITUATION

A-1. In 1982, Israel launched OPERATION PEACE FOR GALILEE designed to destroy the Palestine Liberation Organization (PLO) presence in southern Lebanon. On 1 June, Israeli Defense Forces (IDF) launched a massive assault across the border into southern Lebanon. The Israeli attack focused on the PLO, but the operations quickly involved major ground and air combat between Israel and Syrian forces.

A-2. In the first few weeks, Israeli forces quickly pushed back both the Syrians and the PLO. However, except for some PLO forces isolated in bypassed urban areas, such as Tyre and Sidon, most of the PLO fell back into Beirut (see Figure A-1). By 30 June, Israeli forces had reached the outskirts of southern Beirut, occupied East Beirut, isolated the city from Syria and the rest of Lebanon, and blockaded the sea approaches to the city. Even so, with most of the PLO intact inside and with significant military and political capability, the Israelis had yet to achieve the objective of OPERATION PEACE FOR GALILEE. The Israeli command had to make a decision. It had three choices: permit the PLO to operate in Beirut; execute a potentially costly assault of the PLO in the city; or lay siege to the city and use the siege to successfully achieve the objective. The Israelis opted for the latter.

CONTENTS

ISRAELI MILITARY POSITION

A-3. The Israelis had an excellent position around Beirut. They occupied high ground to the south and west, virtually dominating the entire city. Israeli naval forces controlled the seaward approaches to Beirut. The Israelis' position was also strong defensively, capable of defeating any attempt to break out of or into the city from northern Lebanon or Syria. The Israeli air force had total and complete air superiority. The Israelis controlled the water, fuel, and food sources of West Beirut. Although the PLO forces had stockpiles of food and supplies, the Israelis regulated the food, water, and generating power for the civil population.

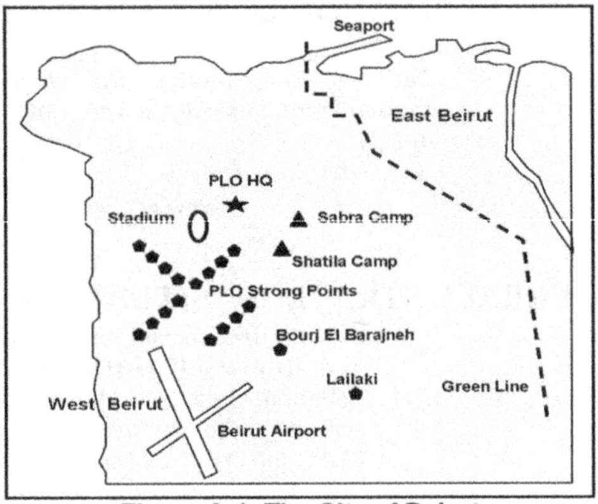

Figure A-1. The City of Beirut

A-4. Despite the superior positioning of Israeli forces, the IDF faced significant challenges to include the combat power of the PLO, Syria, and other threats in Beirut. Israeli doctrine and training did not emphasize urban operations. Additionally, Israel was constrained by its desire to limit collateral damage and friendly and noncombatant casualties. Organizationally, the Israeli army was not optimized to fight in urban terrain. Armor and self-propelled artillery formations dominated the Israeli forces, and most Israeli infantry was mechanized. The Israeli forces had only a few elite formations of traditional dismounted infantry.

PLO MILITARY POSITION

A-5. Despite being surrounded and cut off from support, the PLO position in Beirut offered numerous advantages in addition to the characteristic advantages of urban defense. The PLO had long anticipated an Israeli invasion of southern Lebanon; it had had months to prepare bunkers, obstacles, and the defensive plan of Beirut and other urban areas. Approximately 14,000 Arab combatants in West Beirut readied to withstand the Israeli siege. This was done with the advice of Soviet, Syrian, and east European advisors. The preparation included stockpiling essential supplies in quantities sufficient to withstand a six-month siege. Also, the PLO fighters integrated into the civil populations of the urban areas. Often their families lived with them. The civil population itself was friendly and provided both information and concealment for PLO forces. PLO fighters were experienced in urban combat and knew the urban terrain intimately. PLO forces had been involved in urban fighting against Syrian conventional forces and Christian militias in Beirut several years prior to the Israeli invasion. Finally, the organization of the PLO—

centered on small teams of fighters armed with machine-guns and antitank weapons, and trained in insurgent, hit-and-run tactics—was ideally suited to take maximum advantage of the urban environment.

ROLE OF CIVILIANS

A-6. Various ethnic and religious groups make up the civil population of southern Lebanon. However, West Beirut's population was heavily Palestinian and Lebanese. The civil population of West Beirut was between 350,000 and 500,000. The Palestinian population supported the PLO. The Lebanese population may be described as friendly neutral to the Israelis. Although unhappy under Palestinian dominance, this population was unwilling to actively support Israel. The civilian population was a logistic constraint on the PLO, which would have become significant had the siege lasted longer. The civilians in West Beirut were an even larger constraint on the Israelis. The presence of civilians significantly limited the ability of the Israelis to employ firepower. However, the Palestine refugee camps located in West Beirut were both civilian centers and military bases. The Israeli constraints on artillery and other systems against these parts of the city were much less restrictive than in other parts of West Beirut where the population was mostly Lebanese and where fewer key military targets existed.

A-7. The PLO knew of the Israeli aversion to causing civil casualties and purposely located key military centers, troop concentrations, and logistics and weapons systems in and amongst the population—particularly the refugee Palestinian population in the southern part of West Beirut. Tactically, they used the civilians to hide their forces and infiltrate Israeli positions.

A-8. The friendly Palestinian population provided intelligence to the PLO while the friendly Lebanese population provided intelligence for the IDF. Throughout the siege, the IDF maintained a policy of free passage out of Beirut for all civilians. This policy was strictly enforced and permitted no weapons to leave the city. Some estimates are that as many as 100,000 refugees took advantage of this policy.

INFORMATION OPERATIONS

A-9. The siege of Beirut involved using information operations (IO) to influence the media. PLO information operations were aimed at controlling the media and hence the international perception of the operation. This was done by carefully cultivating a select group of pro-PLO media years before hostilities even began. Once hostilities started, only these media sources were permitted to report from the besieged portions of the city, and they were only shown activities that portrayed the IDF negatively. The IDF did not vigorously counter the PLO plan. In fact, the IDF contributed to it by limiting media access to their activities. The PLO information operations had a successful impact. The international community was constantly pressuring the Israeli government to end hostilities. This put pressure on the IDF to conduct operations rapidly and to limit firepower and casualties.

CONDUCT OF THE URBAN OPERATIONS

A-10. The siege of Beirut began 1 July (see Figure A-2). By 4 July, Israeli forces occupied East Beirut, the Green Line separating East and West Beirut, and dominating positions south of the airport. IDF naval forces also controlled the sea west and north of Beirut. On 3 and 4 July, IDF artillery and naval fire began a regular campaign of firing on military targets throughout West Beirut. On 4 July, the IDF cut power and water to the city.

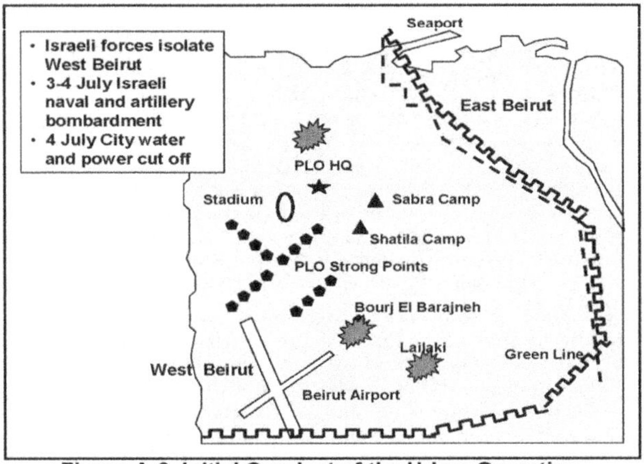

Figure A-2. Initial Conduct of the Urban Operation

A-11. From 5 to 13 July, the Israeli fires continued to pound PLO targets in West Beirut. The PLO gave one significant response, firing on an Israeli position south of the city and causing several casualties. On 7 July, reacting to international pressure, the IDF returned power and water to West Beirut's civil population. On 11 July, the IDF launched its first attack, probing the southern portion of the airport with an armored task force (see Figure A-3). The PLO repulsed this attack and destroyed several IDF armored vehicles.

A-12. On 13 July, both sides entered into a cease-fire that lasted until 21 July. They began negotiations, mediated by international community, to end the siege. The PLO used this period to continue to fortify Beirut. The Israelis used the time to train their infantry and other arms in urban small unit tactics in Damour, a town the Israeli paratroopers had captured.

A-13. The cease-fire ended on 21 July as PLO forces launched three attacks on IDF rear areas. The Israelis responded with renewed and even more vigorous artillery, naval, and air bombardment of PLO positions in the city. The IDF attacks went on without respite until 30 July. On 28 July, the IDF renewed its ground attack in the south around the airport (see Figure A-4). This time IDF forces methodically advanced and captured a few hundred meters of ground establishing a toehold.

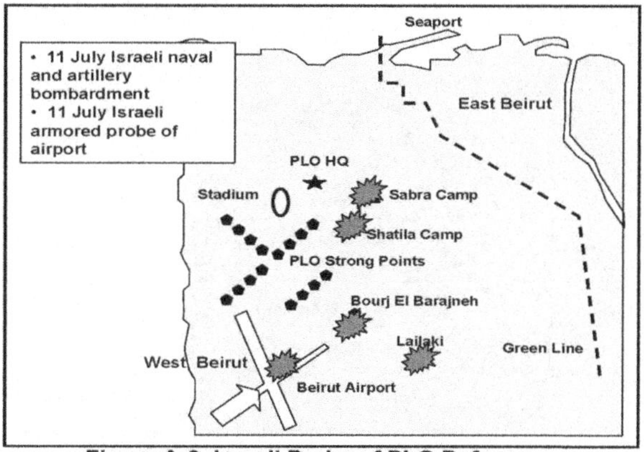

Figure A-3. Israeli Probe of PLO Defenses

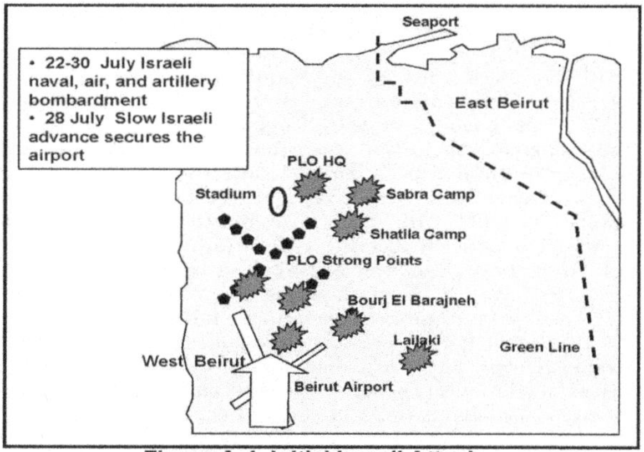

Figure A-4. Initial Israeli Attack

A-14. The Israeli bombardment stopped On 31 July. However, on 1 August the IDF launched its first major ground attack, successfully seizing Beirut airport in the south (see Figure A-5 on page A-6). Israeli armored forces began massing on 2 August along the green line, simultaneously continuing the attack from the south to the outskirts of the Palestinian positions at Ouzai. On 3 August, the Israeli forces continued to reinforce both their southern attack forces and forces along the green line to prepare for continuing offensive operations. On 4 August, the IDF attacked at four different places. This was the much-anticipated major Israeli offensive.

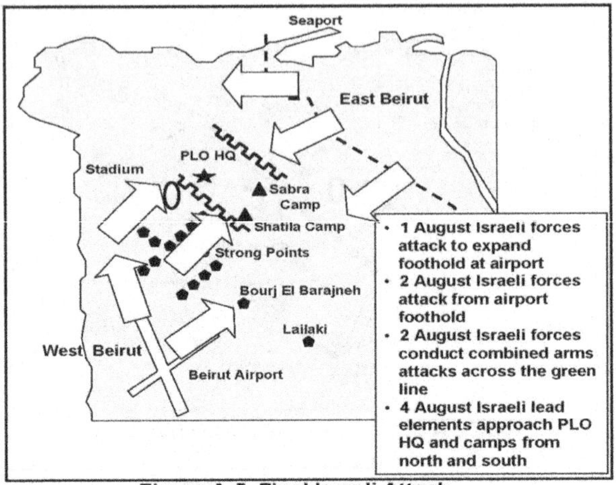

Figure A-5. Final Israeli Attack

A-15. The Israeli attack successfully disrupted the coherence of the PLO defense. The southern attack was the most successful: it pushed PLO forces back to their camps of Sabra and Shatila and threatened to overrun PLO headquarters. Along the green line the IDF attacked across three crossing points. All three attacks made modest gains against stiff resistance. For this day's offensive, the Israelis suffered 19 killed and 84 wounded, the highest single day total of the siege, bringing the total to 318 killed. Following the major attacks on 4 August, Israeli forces paused and, for four days, consolidated their gains and prepared to renew the offensive. Skirmishes and sniping continued, but without significant offensive action. On 9 August, the IDF renewed air and artillery attacks for four days. This activity culminated on 12 August with a massive aerial attack that killed over a hundred and wounded over 400—mostly civilians. A cease-fire started the next day and lasted until the PLO evacuated Beirut on 22 August.

LESSONS

A-16. The Israeli siege of West Beirut was both a military and a political victory. However, the issue was in doubt until the last week of the siege. Military victory was never in question; the issue in doubt was whether the Israeli government could sustain military operations politically in the face of international and domestic opposition. On the other side, the PLO faced whether they could last militarily until a favorable political end could be negotiated. The answer was that the PLO's military situation became untenable before the Israeli political situation did.

A-17. This favorable military and political outcome stemmed from the careful balance of applying military force with political negotiation. The Israelis also balanced the type of tactics they employed against the domestic aversion to major friendly casualties and international concern with collateral damage.

PERFORM FOCUSED INFORMATION OPERATIONS

A-18. The PLO devoted considerable resources and much planning on how to use IO to their best advantage. They chose to focus on media information sources as a means of influencing international and domestic opinion.

A-19. The PLO's carefully orchestrated misinformation and control of the media manipulated international sentiment. The major goal of this effort was to grossly exaggerate the claims of civilian casualties, damage, and number of refugees—and this was successfully accomplished. Actual casualties among the civilians were likely half of what the press reported during the battle. The failure of the IDF to present a believable and accurate account of operations to balance PLO efforts put tremendous pressure on the Israeli government to break off the siege. It was the PLO's primary hope for political victory.

A-20. In contrast to the weak performance in IO, the IDF excelled in psychological operations. IDF psychological operations attacked the morale of the PLO fighter and the Palestinian population. They were designed to wear down the will of the PLO to fight while convincing the PLO that the IDF would go to any extreme to win. Thus defeat was inevitable. The IDF used passive measures, such as leaflet drops and loudspeaker broadcasts. They used naval bombardment to emphasize the totality of the isolation of Beirut. To maintain high levels of stress, to deny sleep, and to emphasize their combat power, the IDF used constant naval, air, and artillery bombardment. They even employed sonic booms from low-flying aircraft to emphasize the IDF's dominance. These efforts helped to convince the PLO that the only alternative to negotiation on Israeli terms was complete destruction.

CONDUCT CLOSE COMBAT

A-21. The ground combat during the siege of Beirut demonstrated that the lessons of tactical ground combat learned in the latter half of the twentieth century were still valid. A small combined arms team built around infantry, but including armor and engineers, was the key to successful tactical combat. Artillery firing in direct fire support of infantry worked effectively as did the Vulcan air defense system. The Israeli tactical plan was sound. The Israelis attacked from multiple directions, segmented West Beirut into pieces, and then destroyed each individually. The plan's success strongly influenced the PLO willingness to negotiate. Tactical patience based on steady though slow progress toward decisive points limited both friendly and noncombatant casualties. In this case, the decisive points were PLO camps, strong points, and the PLO headquarters.

A-22. The willingness to execute close combat demonstrated throughout the siege, but especially in the attacks of 4 August, was decisive. Decisive ground combat was used sparingly, was successful and aimed at decisive points, and was timed carefully to impact on achieving the political objectives in negotiations. The PLO had hoped that their elaborate defensive preparations would have made Israeli assaults so costly as to convince the Israelis not to attack. That the Israelis could successfully attack the urban area convinced the PLO leadership that destruction of their forces was inevitable. For this reason they negotiated a cease-fire and a withdrawal on Israeli terms.

AVOID THE ATTRITION APPROACH

A-23. The Israelis carefully focused their attacks on objects that were decisive and would have the greatest impact on the PLO: the known PLO headquarters and refugee centers. Other areas of West Beirut were essentially ignored. For example, the significant Syrian forces in West Beirut were not the focus of Israeli attention even though they had significant combat power. This allowed the Israelis to focus their combat power on the PLO and limit both friendly casualties and collateral damage.

CONTROL THE ESSENTIAL AND PRESERVE CRITICAL INFRASTRUCTURE

A-24. The Israeli siege assured Israeli control of the essential infrastructure of Beirut. The initial Israeli actions secured East Beirut and the city's water, power, and food supplies. The Israelis also dominated Beirut's international airport, closed all the sea access, and controlled all routes into and out of the city. They controlled and preserved all that was critical to operating the city and this put them in a commanding position when negotiating with the PLO.

MINIMIZE COLLATERAL DAMAGE

A-25. The Israeli army took extraordinary steps to limit collateral damage, preserve critical infrastructure, and put in place stringent rules of engagement (ROE). They avoided randomly using grenades in house clearing, limited the use of massed artillery fires, and maximized the use of precision weapons. With this effort, the Israelis extensively used Maverick missiles because of their precise laser guidance and small warheads.

A-26. The strict ROE, however, conflicted with operational guidance that mandated that Israeli commanders minimize their own casualties and adhere to a rapid timetable. The nature of the environment made fighting slow. The concern for civilian casualties and damage to infrastructure declined as IDF casualties rose. They began to bring more field artillery to bear on Palestinian strong points and increasingly employed close air support. This tension underscores the delicate balance that Army commanders will face between minimizing collateral damage and protecting infrastructure while accomplishing the military objective with the least expenditure of resources—particularly soldiers. ROE is but one tool among many that a commander may employ to adhere to this UO fundamental.

UNDERSTAND THE HUMAN DIMENSION

A-27. The Israelis had a noteworthy (although imperfect and at times flawed) ability to understand the human dimension during their operations against the PLO in Beirut. This was the result of two circumstances. First, the PLO was a threat with which the Israeli forces were familiar after literally decades of conflict. Second, through a close alliance and cooperation with Lebanese militia, the Israelis understood a great deal regarding the attitudes and disposition of the civil population both within and outside Beirut.

SEPARATE NONCOMBATANTS FROM COMBATANTS

A-28. Separating combatants from noncombatants was a difficult but important aspect of the Beirut operation. The Israelis made every effort to positively identify the military nature of all targets. They also operated a free passage system that permitted the passage of all civilians out of the city through Israeli lines. The need to impose cease-fires and open lanes for civilians to escape the fighting slowed IDF operations considerably. Additionally, Israeli assumptions that civilians in urban combat zones would abandon areas where fighting was taking place were incorrect. In many cases, civilians would try to stay in their homes, leaving only after the battle had begun. In contrast, the PLO tied their military operations closely to the civilian community to make targeting difficult. They also abstained from donning uniforms to make individual targeting difficult.

A-29. Earlier in OPERATION PEACE FOR GALILEE when the IDF attacked PLO forces located in Tyre, Israeli psychological operations convinced 30,000 Lebanese noncombatants to abandon their homes and move to beach locations outside the city. However, the IDF was subsequently unable to provide food, water, clothing, shelter, and sanitation for these displaced civilians. IDF commanders compounded the situation by interfering with the efforts by outside relief agencies to aid the displaced population (for fear that the PLO would somehow benefit). Predictably, many civilians tried to return to the city complicating IDF maneuver and targeting—that which the separation was designed to avoid. IDF commanders learned that, while separation is important, they must also adequately plan and prepare for the subsequent control, health, and welfare of the noncombatants they displace.

RESTORE ESSENTIAL SERVICES

A-30. Since essential services were under Israeli command, and had been since the beginning of the siege, the Israelis had the ability to easily restore these resources to West Beirut as soon as they adopted the cease-fire.

TRANSITION CONTROL

A-31. In the rear areas of the Israeli siege positions, the Israeli army immediately handed over civic and police responsibility to civil authorities. This policy of rapid transition to civil control within Israeli lines elevated the requirement for the Israeli army to act as an army of occupation. The Israeli army believed the efficient administration of local government and police and the resulting good will of the population more than compensated for the slightly increased force protection issues and the increased risk of PLO infiltration.

A-32. Upon the cease-fire agreement, Israeli forces withdrew to predetermined positions. International forces under UN control supervised the evacuation of the PLO and Syrian forces from Beirut. These actions were executed according to a meticulous plan developed by the Israeli negotiators and agreed to by the PLO. Israeli forces did not take over and occupy Beirut as a result of the 1982 siege (an occupation did occur later but as a result of changing situations).

SUMMARY

A-33. The Israeli siege of West Beirut demonstrates many of the most demanding challenges of urban combat. In summary, the IDF's successful siege of Beirut emerged from their clearly understanding national strategic objectives and closely coordinating diplomatic efforts with urban military operations. A key part of that synchronization of capabilities was the understanding that the efforts of IDF would be enhanced if they left any escape option open to the PLO. This way out was the PLO's supervised evacuation that occurred after the siege. Although the PLO was not physically destroyed, the evacuation without arms and to different host countries effectively shattered the PLO's military capability. Had Israel insisted on the physical destruction of the PLO in Beirut, it might have failed because that goal may not have been politically obtainable in view of the costs in casualties, collateral damage, and international opinion.

Appendix B

Urban Intelligence Preparation of the Battlefield

Maneuvers that are possible and dispositions that are essential are indelibly written on the ground. Badly off, indeed, is the leader who is unable to read this writing. His lot must inevitably be one of blunder, defeat, and disaster.

Infantry in Battle

The complexity of the urban environment and increased number of variables (and their infinite combinations) increases the difficulty of conducting the intelligence preparation of the battlefield (IPB) for urban operations (UO). Although more intricate, *the IPB process remains essential to the successful conduct of UO.* Conducted effectively, it allows commanders to develop the situational understanding necessary to visualize, describe, and direct subordinates in successfully accomplishing the mission.

URBANIZATION OF IPB

B-1. IPB is a systematic process for analyzing the environment and the threat in a specific geographic area—the area of operations (AO) and its associated area of interest. (The area of interest might include areas that are not contiguous with the AO.) It provides direction for the intelligence system, drives the military decision-making process, and supports targeting and battle damage assessment (see Figure B-1). The procedure (as

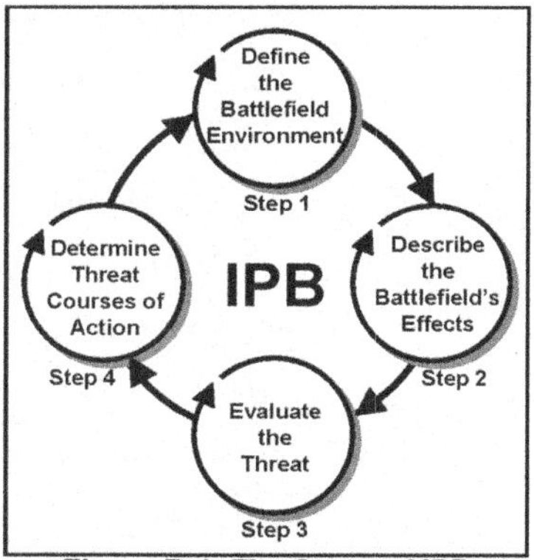

Figure B-1. The Steps of IPB

CONTENTS

well as each of its four steps) is performed continuously throughout the planning, preparation, and execution of an urban operation.

UNAFFECTED PROCESS

B-2. The IPB process is useful at all echelons and remains constant regardless of operation or environment. However, urban IPB stresses some aspects not normally emphasized for IPBs conducted for operations elsewhere. The complex mosaic is comprised of the societal, cultural, or civil dimension of the urban environment; the overlapping and interdependent nature of the urban infrastructure; and the multidimensional terrain. This mosaic challenges the conduct of urban IPB. There is potential for the full range of Army operations to be executed near-simultaneously as part of a single major operation occurring in one urban area with the multiple transitions. Such precision stresses the importance of a thorough, non-stop IPB cycle aggressively led by the commander and executed by the entire staff. Overall, the art of applying IPB to UO is in properly applying the steps to the specific environment and threat. In UO, this translates to understanding and analyzing the significant characteristics of the environment and the role that its populace has in threat evaluation. FM 34-130 details how to conduct IPB; FM 34-3 has the processes and procedures for producing all-source intelligence. This appendix supplements the information found there; it does not replace it.

INCREASED COMPLEXITY

B-3. Uncovering intricate relationships takes time, careful analysis, and constant refinement to determine actual effects on friendly and threat courses of action (COAs). These relationships exist among—

precisely because it is least likely and, therefore, may be intended to maximize surprise. Misunderstanding the urban environment's effect on potential friendly and threat COAs may rapidly lead to mission failure and the unnecessary loss of soldiers' lives and other resources.

Training, Experience, and Functional Area Expertise

B-5. Not all information about the urban environment is relevant to the situation and mission—hence the difficulty. Although it may appear daunting, institutional education, unit training, and experience at conducting an urban IPB will improve the ability to rapidly sort through all the potential information to separate the relevant from merely informative. (This applies to any new or difficult task.) The involvement and expertise of the entire staff will allow commanders to quickly identify the important elements of the environment affecting their operations. Fortunately, IPB is a methodology comprehensive enough to manage the seemingly overwhelming amounts of information coming from many sources.

B-6. As in any operational environment, tension exists between the desire to be methodical and the need to create the tempo necessary to seize, retain, and exploit the initiative necessary for decisive UO. Quickly defining the significant characteristics of the urban environment requiring in-depth evaluation (not only what we need to know but what is possible to know) allows rapid identification of intelligence gaps (what we know versus what we don't know). Such identification leads to priority information requirements (PIR) and will drive the intelligence, surveillance, and reconnaissance (ISR) plan (how will we get the information we need). FM 3-55 and Chapter 4 discuss ISR. Commanders carefully consider how to develop *focused* PIR to enable collectors to more easily weed relevant information from the plethora of information. Commanders can make better decisions and implement them faster than a threat can react.

AMPLIFIED IMPORTANCE OF CIVIL (SOCIETAL) CONSIDERATIONS

B-7. The Army focuses on warfighting. The experiences in urban operations gained at lower echelons often center on the tactics of urban offensive and defensive operations where the influences of terrain and enemy frequently dominate. At higher echelons, the terrain and enemy are still essential considerations, but the societal component of the urban environment is more closely considered. Moreover, the human or civil considerations gain importance in support operations and stability operations regardless of the echelon or level of command. In addition to the echelon and the type of operation, a similar relationship exists between the key elements of the urban environment and other situational factors. These factors can include where the operation lies within the spectrum of conflict or the level of war and the conventional or unconventional nature of the opposing threat. Figure B-2 on page B-4 graphically represents the varying significance of these elements to urban IPB. Population effects are significant only in how they affect the threat, Army forces, and overall mission accomplishment.

B-8. Describing the battlefield's effects—step two of the IPB—ascribes meaning to the characteristics analyzed. It helps commanders understand how the environment enhances or degrades friendly and threat forces and capabilities. It also helps commanders understand how the environment supports the population. It also explains how *changes* in the "normal" urban environment (intentional or unintentional and because of threat or friendly activities) may affect the population. Included in this assessment are matters of perception. At each step of the IPB process, commanders try to determine the urban society's perceptions of ongoing activities to ensure Army operations are viewed as intended. Throughout this process, commanders, staffs, and analysts cannot allow their biases—cultural, organizational, personal, or cognitive—to markedly influence or alter their assessment (see FM 34-3). This particularly applies when they analyze the societal aspect of the urban environment. With so many potential groups and varied interests in such a limited area, misperception is always a risk.

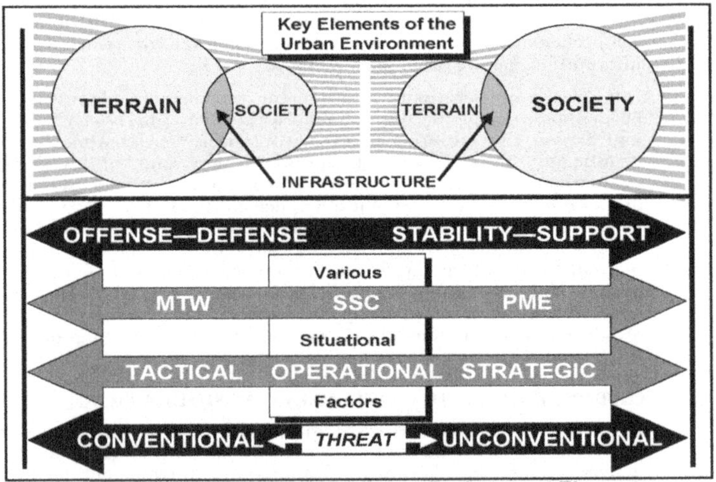

Figure B-2. Relevance of Key Urban Environment Elements

SIGNIFICANT CHARACTERISTICS

B-9. For IPB to remain effective in UO, its analysis must include the urban environment's attributes—man-made terrain, society, and infrastructure—and an evaluation of characteristics traditionally included in IPB: the underlying natural terrain (to include weather) and the threat. Because the urban environment is so complex, it is useful to break it into categories. Then commanders can understand the intricacies of the environment that may affect their operations and assimilate this information into clear mental images. Commanders can then synthesize these images of the environment with the current status of friendly and threat forces and develop a desired end state. Then they can determine the most decisive sequence of activities that will move their forces from the current state to the end state. Identifying and understanding the environment's characteristics (from a friendly, threat, and noncombatant perspective) allows commanders to establish and maintain situational understanding. Then they can develop appropriate COAs and rules of engagement that will lead to decisive mission accomplishment.

B-10. Figures B-3, B-4, and B-5 are not intended to be all-encompassing lists of urban characteristics. They provide a starting point or outline useful for conducting an urban IPB that can be modified to meet the commander's requirements. Commanders and staffs can compare the categories presented with those in the civil affairs area study and assessment format found in FM 41-10 and the IPB considerations for stability operations and support operations found in FM 34-7.

MULTIPLE OVERLAPS

B-11. Since the urban environment is comprised of a "system of systems," considerations among the key elements of the environment will overlap during urban IPB analysis. For example, boundaries, regions, or areas relate to a physical location on the ground. Hence, they have urban *terrain* implications. These boundaries, regions, or areas often stem from some historical, religious, political, or social aspect that could also be considered a characteristic of the urban *society*. Overlaps can also occur in a specific category, such as *infrastructure*. For instance, dams are a consideration for their potential effects on transportation and distribution (mobility), administration and human services (water supply), and energy (hydroelectric).

B-12. This overlap recognition is a critical concern for commanders and their staffs. In "taking apart" the urban environment and analyzing the pieces, commanders and staffs cannot lose perspective of how each piece interacts with any other and as part of the whole. Otherwise, their vision will be shortsighted, and they will fail to recognize the second- and third-order effects of their proposed COAs; the actual end state differing dramatically from the one envisioned by the commander. The increased density of combatants and noncombatants, infrastructure, and complex terrain means that a given action will likely have unintended consequences. Those consequences will be more widely felt and their impact will spread in less time than in other environments. These unintended results may have important strategic and operational consequences. The multiple ways these dynamic urban elements and characteristics combine make it necessary to approach each urban environment as a unique IPB challenge.

URBAN TERRAIN AND WEATHER

Terrain

B-13. Earlier admonitions that civil considerations are more closely considered in UO do not necessarily mean that consideration for urban terrain is de-emphasized. In every urban operation, terrain and its effects on both threat and friendly forces is assessed and understood. Then commanders can quickly choose and exploit the terrain (and weather conditions) that best supports their missions. Terrain analysis thoroughly assesses urban structures as well as the ground on which they stand (see Figure B-3 and FM 5-33). An analysis of urban terrain first considers broader urban characteristics and effects and progress to a more detailed examination.

Forms and Functions

divides the two resident ethnic groups: Albanians and Serbs. The separation became significant at both the strategic and tactical levels during 1999 deployments to Kosovo. Army forces had to ensure that the Orthodox Church located south of the Ibar was accessible to Serbs residing in the north. North Atlantic Treaty Organization (NATO) peacekeepers built a footbridge across the river that allowed reliable, safe passage. The natural feature separating the two groups assisted NATO troops in maintaining stability in the region.

B-15. **Man-Made Terrain.** Building composition, frontages, placement, forms and functions, size, floor plans, and window and door locations affect maneuver, force positioning, and weapons deployment considerations. Angles, displacement, surface reflection, and antenna locations influence command and control. Structures also influence ISR operations. The increased density and volume created by man-made structures increase how much information commanders and their staffs collect and assess as well as the number of forces required. Building materials and construction will also influence force structures to include weapons and equipment required. The ability to maneuver through the urban dimensions—airspace, supersurface (including intrasurface), surface, and subsurface—and shoot through walls, ceilings, and floors also creates increased psychological stress. The characteristics of man-made terrain can also be analyzed using OAKOC.

Weather

B-16. Weather and its effects are often considered when examining the military aspects of terrain. Military aspects of weather include temperature (heat and cold), light conditions, precipitation (cloud cover, rain, snow, fog, and smog), and wind. Their military effects during UO are similar to any operational environment (see FM 34-81 and FM 34-81-1). Extremes of heat and cold affect weapon systems and the soldiers that operate them. The extra luminescence provided by the ambient light of an urban area, unless controlled, may affect night vision capabilities and the ability of the Army to "own the night." Precipitation affects mobility and visibility. Smog inversion layers are common over cities. An inversion layer may trap smoke and chemicals in the air to the detriment of soldiers' health. (If the conditions are severe enough, it might require the use of protective masks.) Winds, which may increase as they are funneled through urban canyons, may—

medical, and rescue). Heavy rains may have similar effects on poorly

History (continued)

support the overall urban operation with the level of detail that commanders and their staffs can achieve. However, an IPB that fails to devote enough time and resources to societal analysis can find large elements of the population turned against the Army force. Analyzing the urban society first may help to focus or limit further analysis of the terrain and infrastructure, saving time and ISR resources.

URBAN INFRASTRUCTURE

B-19. Functional and analytical overlap readily appears when examining urban infrastructures (see Figure B-5). They are composed of physical structures or facilities and people. Hence, much of the analysis conducted for terrain and society can apply when assessing the urban infrastructure. For example, commanders, staffs, and analysts could not effectively assess the urban economic and commercial infrastructure without simultaneously considering labor. All aspects of the society relate and can be used to further analyze the urban work force since they are a subelement of the urban society. Similarly, the OAKOC aspects used to evaluate terrain may also apply to the urban infrastructure, especially considerations of key terrain.

Transportation and Distribution

Administration and Human Services

B-21. A general study of guerrilla and insurgent tactics, techniques, and procedures may prove beneficial to many types of operations regardless of the actual composition or type of threat forces. Insurgent strategies and tactics may work especially well in this complex environment and will likely be a part of any threat COA. Particularly, commanders understand how a threat might restrict itself by the laws of land warfare and similar conventions, or exploit the use of these conventions to its own gain. Commanders can refer to FM 31-20-3 for more information. For many of the above threats, no doctrinal templates may exist. Commanders, staffs, and analysts evaluate, update (or create), and manage threat databases early (and continuously) in the IPB process.

> *. . . [T]he adversaries of freedom . . . send arms, agitators, aid, technicians and propaganda to every troubled area. . . . [S]ubversives and saboteurs and insurrectionists . . . [possess] the power to conscript talent and manpower for any purpose, . . . and long experience in the techniques of violence and subversion. . . . It is a contest of will and purpose as well as force and violence—a battle for minds and souls as well as lives and territory.*
>
> John F. Kennedy
> Message to Congress, 27 April 1961

ENVIRONMENTAL THREATS

B-22. While threats vary, they share a common characteristic: the capability and intent to conduct violence against Army forces to negatively influence mission accomplishment. These threats are often the most recognizable for forces trained for warfighting—these are often the enemy. Army units be able to conduct full spectrum operations—offense, defense, stability, and support. Commanders broaden their concept of the threat when analyzing the urban environment's terrain, societal, and infrastructure characteristics. This analysis includes many environmental dangers (potentially affecting both sides of a conflict as well as noncombatants) such as—

CIVILIANS

B-23. In a major theater war where offensive and defensive operations are conducted against a clear enemy, the third step of IPB—identify the threat—is readily accomplished. Its well-established procedures include updating or creating threat models and identifying threat capabilities. This same analytic process includes modeling population subgroups. The process applies to smaller-scale contingencies, peacetime military engagement activities where stability operations and support operations dominate, and urban offensive and defensive operations where civilians are in close proximity to Army forces. This adaptation is necessary to further broaden the threat concept to include specific elements of the urban society and, in some instances, non-governmental organizations (NGOs) and other civilian agencies working in the urban area. In many stability operations, this modification can account for opposing armed forces that are not an enemy but are a threat to the mission. As discussed in Chapter 8, Army forces in many stability operations and support operations must avoid classifying or thinking of these threats as *the enemy*.

Need for More Accurate Categories

B-24. Army forces recognized that the threat evaluation was not a straightforward assessment of the capabilities of a known, armed enemy. This resulted in developing categories for assessing the disposition of population subgroups within an AO: enemy, neutral, or friendly. Sectors of the population were labeled based on which side (if any) each group seemed to favor. This early method helped to mitigate Army forces' situational uncertainty. It provides a general idea of the level of support or resistance Army forces might expect by elements of the urban population.

B-25. Aside from the flawed labeling of every threat as an enemy, the initial attempt at categorization was a good first step. However, it required refinement to better indicate the level of threat or utility that civilian groups pose for Army forces conducting UO. It also provides a clearer basis for detecting and monitoring shifts in key or relevant relationships (see Figure B-6). Commanders note that where a group or subgroup falls along this continuum is relative to the perspective from which it is viewed. This is an especially important consideration in multinational and interagency UO. During operations in Somalia, US Army forces may have viewed a particular clan as a hostile element. The United Nations' Italian contingent, with their colonial background in the region, may have considered the same clan as neutral or even an ally.

B-26. Although necessary and greatly improved, commanders recognize that no system of categorization will precisely classify any given group; no system can reflect the overall nature and complexity of the urban society. A single group may fit in a particular category. It may also have components in two or more categories simultaneously. Often, it can shift among categories during an operation. A given group may have individuals in it who have interests identical to or different from that group and these individual interests may change over time.

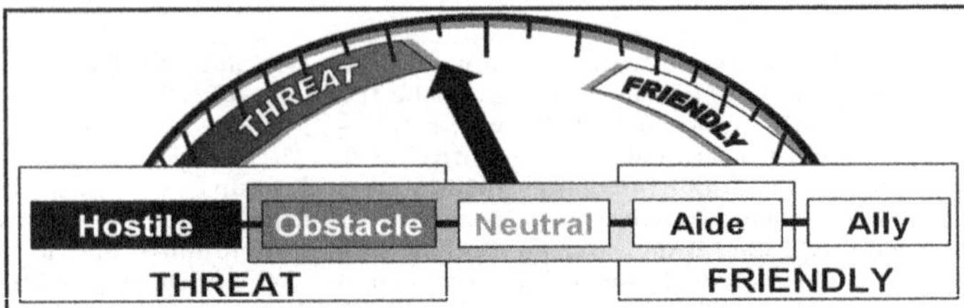

Hostile:	An element (*individual, group, or organization*) that has the intent to negatively influence mission accomplishment by supporting enemy forces (or another hostile element) or exploiting friendly vulnerabilities but is not engaging in protracted combat operations.
Obstacle:	An element whose presence or activities serve to unintentionally impede, disrupt, or constrain friendly operations by imposing additional losses in personnel, time, equipment, and resources. Its interests may or may not be compatible with friendly objectives.
Neutral:	An element whose presence, activities, and overall interests do not conflict with friendly, enemy, or hostile forces.
Aide:	An element whose *present* intentions are—either wittingly or unwittingly—compatible with friendly force objectives.
Ally:	An element whose *fundamental* interests and intent align with friendly force objectives.

Figure B-6. Continuum of Relative Interests

B-27. A peace enforcement operation illustrates the varying nature of groups. An identified criminal group might be classified as an *obstacle* to the commander's mission because its illegal activities impede unit progress. Its compelling interest, however, is to make money rather than interfere with friendly forces. In the same operation, one of the armed belligerents may be intent on disrupting the peace process and would be, therefore, classified as a *hostile*. (Again, not as an enemy unless they engaged in prolonged combat operations against the peacekeeping force.) The belligerent force may finance the criminal organization to assist in further obstructing the peace mission. The criminal organization moves from being an obstacle to that of a hostile.

Shifting Civilian Interests and Intent

Among other applicable lessons (see also the vignette in Chapter 8), the Los Angeles riots of 1992 illustrate how urban population groups can shift their relative positions due to changing conditions in an urban AO. Several gangs exist in the Los Angeles area. Usually, these gangs are hostile to one another. During the riots, however, several rival gangs formed a "united front" against what was seen as a larger obstacle to their own interests: law enforcement. As a result, the hostile gangs became one another's aides during that time.

B-28. This classification effort, therefore, is not a one-time undertaking; commanders constantly review and update it (like the entire IPB process itself). Groups or individuals can be influenced into assisting either the friendly or opposing force. People will also act opportunistically, shifting support and alliances as perceived advantages arise. Even seemingly passive and law-abiding members of the urban society may conduct themselves in unexpected ways given the right conditions (mob violence, for example).

Similarities, Differences, Capabilities, and Vulnerabilities

B-29. **Focal Points for Analysis.** Similarities and differences in interests and interdependencies between groups are often focal points for analysis (and the allocation of ISR assets). They may indicate how commanders may influence, coerce, or align civilian interests and intentions with mission objectives. Simultaneously, commanders consider an analysis of the civilian element's (individual, group, or organization) *capability* to influence the accomplishment of friendly objectives. They also consider civilians' vulnerabilities and dependencies. If a civilian group's fundamental interests align with friendly objectives and this group has the intent to assist friendly forces, it is clearly an ally. However, with limited or no capability, a specific group will not help move the commander any closer to achieving his desired end state and accomplishing the mission. Then the commander would normally limit the resources expended on this group to those necessary for maintaining their commitment to common goals and objectives.

B-30. **Creating Civilian Capability as Aide or Ally.** In contrast, commanders may provide a group with resources to enhance or create the capability to assist in mission accomplishment. They may do this if they felt that the assistance gained (or reduction to threat support) exceeds the potential diminishment of their own force's capabilities from losing those same resources. Commanders would also consider the group's dependencies (such as food, infrastructure, and protection) and overall vulnerability to outside influence. If vulnerable to friendly influence or control (understanding urban societal considerations and matters of perception are critical in this regard), then forces are likely susceptible to enemy or hostile manipulation. Even if commanders can generate extra resources (and not significantly affect their own force's capabilities), they still conduct this same cost-benefit analysis to determine which civilian group (if any) should receive resources.

Greatest Potential Nearer the Center

B-31. The most critical population sectors often are those nearest the middle of the spectrum, particularly if their capabilities (or potential) significantly degrades or enhances mission accomplishment. These are the obstacle, neutral, or aide categories. If their interests can be adequately understood, commanders may have great chance to influence the population segment and significantly contribute to mission accomplishment.

Political and Strategic Concerns

B-32. The aide category may be of significant political or strategic concern. An aide group may be invaluable for accomplishing intermediate objectives but become a vulnerability to accomplishing a larger stability operation. (Even an urban offensive or defensive operation is likely to transition to a stability operation.) Commanders may provide resources to a criminal organization to assist in defeating insurgent forces during urban combat operations. Once these forces have been subdued, the interests (monetary gain and defeat of the threat) linking friendly forces with this criminal element disappear. What may remain is a criminal organization with more power than a reconstituted or newly established law enforcement agency and a truly destabilizing force. This also illustrates the second- and third-order thinking that will be required of commanders and their staffs during UO.

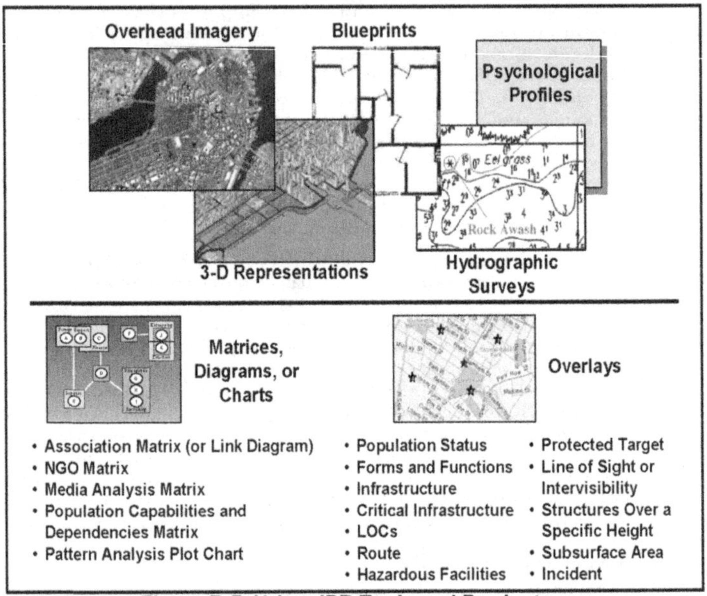

Figure B-7. Urban IPB Tools and Products

URBAN IPB TOOLS AND PRODUCTS

B-33. Adapting IPB to UO involves recognizing the intent of each of the steps of the process and adapting analytic tools and products to help meet those intentions in a complex environment (see Figure B-7 and FM 3-34.230). Standard tools and products include: modified combined obstacle overlays and doctrinal, situation, event, and decision support templates or matrices. In addition to these standards aids, staffs and analysts may develop or produce other innovative tools to assist commanders in their situational understanding of the complex urban environment. Staffs and analysts may also initiate requests for products (or information) from their higher headquarters or other agencies with the technical means or control over assets when the capability lies outside the Army force's means. The tools that developed or requested may include—

OVERHEAD IMAGERY

B-34. Recent satellite imagery or aerial photography will be required for most types of UO. Such images clarify vague and inaccurate maps and other graphic representations. Satellite assets provide responsive data input into the geographic information systems (GIS). (The National Imagery and Mapping Agency [NIMA] and other intelligence sources prepare data sets.) GIS will often form the basis for creating the three-dimensional representations and the various overlays described below. Frequently updated (or continuous real-time) satellite or aerial imagery may be required for detailed pattern analysis and maintaining accurate situational understanding. For example,. imagery taken during an area's rainy season may appear significantly altered during the summer months.

THREE-DIMENSIONAL REPRESENTATIONS

B-35. Often, physical or computer-generated (virtual) three-dimensional representations may be required to achieve situational understanding. These representations include specific sections of the urban area or specific buildings or structures. Such detail is particularly important for special operating forces and tactical-level units. These units require detail to achieve precision, increase the speed of the operation, and lessen friendly casualties and collateral damage.

INFRASTRUCTURE BLUEPRINTS

B-36. Urban police, fire, health, public utilities, city engineers, realtors, and tourist agencies often maintain current blueprints and detailed maps. Such documents may prove useful to update or supplement military maps or to clarify the intricacies of a specific infrastructure. They may prove critical in operations that require detailed information to achieve the speed and precision required for success. Without such detail, analysts determine interior configurations based on a building's outward appearance. That task becomes more difficult as the building size increases.

HYDROGRAPHIC SURVEYS

B-37. Many urban areas are located along the world's littorals regions and major rivers. Therefore, commanders may need hydrographic surveys to support amphibious, river crossing, and logistic operations.

PSYCHOLOGICAL PROFILES

B-38. Psychological profiles analyze how key groups, leaders, or decision-makers think or act—their attitudes, opinions, and views. They include an analysis of doctrine and strategy, culture, and historical patterns of behavior. The degree to which the attitudes, beliefs, and backgrounds of the military either reflect or conflict with the urban populace's (or civilian leadership's) core values is extremely important in this analysis. Psychological profiles help to assess the relative probability of a threat (or noncombatant group) adopting various COAs as well as evaluating a threat's vulnerability to deception. These profiles are derived from open-source intelligence as well as signals and human intelligence.

MATRICES, DIAGRAMS, OR CHARTS

B-39. Matrices, diagrams, and charts help to identify key relationships among friendly and threat forces and other significant elements of the urban environment. These tools and products include—

outlying high-rise areas, commercial ribbon areas, and residential areas, to include shantytowns.

depression capabilities of weapon systems and potential threat ambush locations—again, affecting maneuver options.

Appendix C

Operations in Somalia: Applying the Urban Operational Framework to Support and Stability

It's impossible for an American mother to believe that a Somali mother would raise children to avenge the clan.

Major General Thomas M. Montgomery

GENERAL SITUATION

C-1. Following decades of political unrest and the fall of Somali dictator Siad Barre, a civil war broke out as 14 clans vied for power. The resulting nation composed of hostile social factions was held together by weak political alliances—none strong enough to unite and lead the country to national reconciliation. An ongoing drought led to famine and compounded the ethnic tensions and political instability. This volatile situation rapidly led to a phased US involvement (see Figure C-1). Army forces combined, sequenced, and proportionally emphasized the different types of operations to accomplish

Operation	Dates	UN Security Council Resolution	Relative Proportionality Between Types of Operations
PROVIDE RELIEF (UNOSOM I)	Aug 92 – Dec 92	UNSCR# 751 24 Apr 92	Support, Stability, Defense, Offense
RESTORE HOPE (UNITAF)	Dec 92 – May 93	UNSCR# 794 3 Dec 92	Support, Stability, Defense, Offense
CONTINUED HOPE (UNOSOM II)	May 93 – Mar 94	UNSCR# 814 26 Mar 93	Stability, Support, Offense, Defense

Figure C-1. Phases of US Involvement in Somalia

CONTENTS

changing political objectives. Throughout all operations in Somalia, urban areas were critical to achieving mission success.

SOMALI OPERATIONS

INITIAL UN RESPONSE

C-2. The United Nations (UN) initially responded to requests for assistance from international relief organizations by sending supplies and other forms of humanitarian aid to Somalia. However, widespread looting, fighting between gangs, and other lawlessness prevented supplies from reaching the hungry and sick. Only 20 percent of the food entering the country reached the people who needed it. An estimated 25 percent of Somalia's 6 million people died of starvation or disease. In April 1992, the UN issued Security Council Resolution (UNSCR) 751 which authorized 50 *unarmed* observers, but the action had little effect. Under continuing pressure for additional measures to ensure the delivery of supplies and relief, the UN authorized 500 *armed* peacekeepers (furnished by Pakistan and transported by US sea- and airlift) to protect humanitarian workers. The battalion's limited mission, designated UN Operations in Somalia I (UNOSOM I), encompassed safeguarding the unloading of ships and providing convoy security.

PROVIDE RELIEF (UNOSOM I)

C-3. In July 1992, the UN requested an increased airlift of supplies and the US quickly responded. US Central Command (CENTCOM) activated joint task force (JTF) OPERATION PROVIDE RELIEF. Based on careful mission analysis, CENTCOM limited the JTF's actions to—

humanitarian relief operations in Somali" and demanded "all factions in Somalia immediately cease hostilities." To allay concerns of colonialism by a number of African countries, the UN Secretary-General was given oversight of the operation. The resolution also required soldiers to be withdrawn once order was restored; however, it provided no exit strategy. As clearly as possible, the CENTCOM mission statement for OPERATION RESTORE HOPE reflected the UN mandate:

> When directed by the [President or the Secretary of Defense], USCINCCENT will conduct joint/combined military operations in Somalia to secure the major air and sea ports, key installations and food distribution points, to provide open and free passage of relief supplies, provide security for convoys and relief organization operations, and assist UN/NGOs in providing humanitarian relief under UN auspices. Upon establishing a secure environment for uninterrupted relief operations, USCINCCENT terminates and transfers relief operations to UN peacekeeping forces.

C-5. Mogadishu was the largest port in the country and the focal point of previous humanitarian relief activities of nongovernmental organizations (NGOs). It was also the headquarters of the coalition of 20 nations and over 30 active humanitarian relief organizations. As such, Mogadishu became the entry point for the operational buildup of the multinational force known as Unified Task Force (UNITAF) and the key logistic hub for all operations in Somalia. UNITAF immediately gained control over the flow of relief supplies into and through Mogadishu and stabilized the conflict among the clans. In less than a month, UNITAF forces expanded control over additional ports and interior airfields. They secured additional distribution sites in other key urban areas in the famine belt to include Baidoa, Baledogle, Gialalassi, Bardera, Belet Uen, Oddur, Marka, and the southern town of Kismayo (see Figure C-2). With minimal force, the US-led UNITAF established a secure environment that allowed relief to reach those in need, successfully fulfilling its limited—yet focused—mandate.

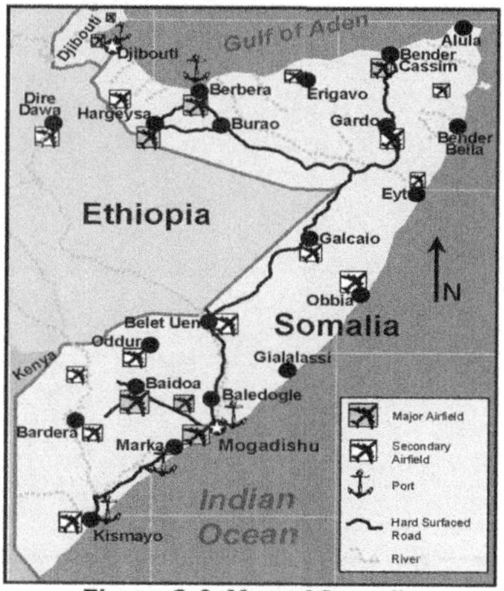

Figure C-2. Map of Somalia

CONTINUED HOPE (UNOSOM II)

C-6. In March 1993, the UN issued UNSCR 814 establishing a permanent peacekeeping force, UNOSOM II. However, the orderly transition from UNITAF to UNOSOM II was repeatedly delayed until May 1993. (The UN

Secretary-General urged the delay so that US forces could effectively disarm bandits and rival clan factions in Somalia.) This resolution was significant in two critical aspects:

ASSESS

C-10. Although accomplished to varying degrees, US forces failed to adequately assess the urban environment, especially the society. Somali culture stresses the unity of the clan; alliances are made with other clans only when necessary to elicit some gain. Weapons, overt aggressiveness, and an unusual willingness to accept casualties are intrinsic parts of the Somali culture. Women and children are considered part of the clan's order of battle.

C-11. Early in the planning for OPERATION RESTORE HOPE, US forces did recognize the limited transportation and distribution infrastructure in Mogadishu. The most notable was the limited or poor airport and harbor facilities and its impact on the ability of military forces and organizations to provide relief. Therefore, a naval construction battalion made major improvements in roads, warehouses, and other facilities that allowed more personnel, supplies, and equipment to join the relief effort faster.

UNDERSTANDING THE CLAN (THE HUMAN DIMENSION)

C-12. During OPERATION RESTORE HOPE, the UNITAF worked with the various clan leaders as the only recognized leadership remaining in the country. The UNITAF was under the leadership of LTG Robert B. Johnston and US Ambassador to Somalia, Robert Oakley. In addition, UNITAF forces also tried to reestablish elements of the Somali National Police—one of the last respected institutions in the country that was not clan-based. This reinstated police force manned checkpoints throughout Mogadishu and provided crowd control at feeding centers. Largely because of this engagement strategy, the UNITAF succeeded in its missions of stabilizing the security situation and facilitating humanitarian relief. Before its termination, the UNITAF also worked with the 14 major Somali factions to agree to a plan for a transitional or transnational government.

C-13. The UN Special Representative of the Secretary-General, retired US Navy admiral Jonathon Howe, worked with the UNOSOM II commander, Turkish General Cevik Bir. During OPERATION CONTINUED HOPE, Howe and General Bir adopted a philosophy and operational strategy dissimilar to their UNITAF predecessors. Instead of engaging the clan leaders, Howe attempted to marginalize and isolate them. Howe initially attempted to ignore Aideed and other clan leaders in an attempt to decrease the warlord's power. Disregarding the long-established Somali cultural order, the UN felt that, in the interest of creating a representative, democratic Somali government, they would be better served by excluding the clan leadership. This decision ultimately set the stage for strategic failure.

THREAT STRATEGY AND TACTICS

C-14. During OPERATION RESTORE HOPE, US forces also failed to properly analyze their identified threat's intent and the impact that the urban environment would have on his strategy, operations, and tactics. The UN began to view eliminating Aideed's influence as a decisive point when creating an environment conducive to long-term conflict resolution. Aideed's objective, however, remained to consolidate control of the Somali nation under his leadership—his own brand of conflict resolution. He viewed the UN's operational center of gravity as the well-trained and technologically advanced American military forces, which he could not attack directly. He identified a potential American vulnerability—the inability to accept casualties for an operation not vital to national interests—since most Americans still viewed Somalia as a humanitarian effort. If he could convince the American public that the price for keeping troops in Somalia would be costly, or that their forces were hurting as many Somalis as they were helping, he believed they would withdraw their forces. If US forces left, the powerless UN would leave soon after, allowing Aideed to consolidate Somalia under his leadership.

VULNERABILITY AND RISK ASSESSMENT

C-15. US forces failed to assess and anticipate that Aideed would adopt this asymmetric approach and attack the American public's desire to remain involved in Somalia. By drawing US forces into an urban fight on his home turf in Mogadishu, he could employ guerrilla insurgency tactics and use the urban area's noncombatants and its confining nature. Such tactics made it difficult for the US forces to employ their technological superiority. If US forces were unwilling to risk harming civilians, his forces could inflict heavy casualties on them, thereby degrading US public support for operations in Somalia. If, on the other hand, the US forces were willing to risk increased civilian casualties to protect themselves, those casualties would likely have the same effect.

C-16. However, an assessment of the Somali culture and society should have recognized the potential for Aideed's forces to use women and children as cover and concealment. Accordingly, the plan should have avoided entering the densely populated Bakara market district with such restrictive rules of engagement. As legitimacy is critical to stability operations, TF Ranger should have been prepared and authorized to employ nonlethal weapons, to include riot control gas, as an alternative to killing civilians or dying themselves.

C-17. US forces also failed to assess and recognize the critical vulnerability of their helicopters in an urban environment and the potential impact on their operations. TF Ranger underestimated the threat's ability to shoot down its helicopters even though they knew Somalis had attempted to use massed rocket-propelled grenade (RPG) fires during earlier raids. (Aideed brought in fundamentalist Islamic soldiers from Sudan, experienced in downing Russian helicopters in Afghanistan, to train his men in RPG firing techniques). In fact, the Somalis had succeeded in shooting down a UH-60 flying at rooftop level at night just one week prior to the battle. Instead, TF Ranger kept their most vulnerable helicopters, the MH-60 Blackhawks, loitering for forty minutes over the target area in an orbit that was well within Somali RPG range. The more maneuverable AH-6s and MH-6s could have provided the necessary fire support. Planning should have included a ready ground reaction force, properly task organized, for a downed helicopter contingency.

C-18. Information operations considerations apply throughout the entire urban operational framework; however, operations security (OPSEC) is critical to both assessment and shaping. OPSEC requires continuous assessment throughout the urban operation particularly as it transitions among the range of military operations and across the spectrum of conflict. As offensive operations grew during OPERATION CONTINUED HOPE, US forces did little to protect essential elements of friendly information. Combined with the vulnerability of US helicopters, Aideed's followers used US forces' inattention to OPSEC measures to their advantage. The US base in Mogadishu was open to public view and Somali contractors often moved about freely. Somalis had a clear view both day and night of the soldiers' billets. Whenever TF Ranger would prepare for a mission, the word rapidly spread through the city. On 3 October 1993, Aideed's followers immediately knew that aircraft had taken off and, based on their pattern analysis of TF Ranger's previous raids, RPG teams rushed to the rooftops along the flight paths of the task force's Blackhawks.

SHAPE

C-19. One of the most critical urban shaping operations is isolation. During OPERATION CONTINUED HOPE, US forces largely discounted other essential elements of friendly information and did not establish significant public affairs and psychological operations (PSYOP) initiatives. In fact, Army forces lacked a public affairs organization altogether. Consequently, Aideed was not isolated from the support of the Somali people. This failure to shape the perceptions of the civilian populace coupled with the increased use of lethal force (discussed below) allowed Aideed to retain or create a sense of legitimacy and popular support.

C-20. During OPERATION RESTORE HOPE, Aideed conducted his own PSYOP efforts through "Radio Aideed"—his own radio station. UNITAF countered these efforts with radio broadcasts. This technique proved so effective that Aideed called MG Anthony C. Zinni, UNITAF's director of operations, over to his house on several occasions to complain about UNITAF radio broadcasts. General Zinni responded, "if he didn't like what we said on the radio station, he ought to think about his radio station and we could mutually agree to lower the rhetoric." This approach worked.

DOMINATE

C-21. The complexity of urban operations requires unity of command to identify and effectively strike the center of gravity with overwhelming combat power or capabilities. Complex command and control relationships will only add to the complexity and inhibit a commander's ability to dominate and apply available combat power to accomplish assigned objectives. Stability operations and support operations as seen in Somalia required commanders to dominate only within their supporting role and, throughout, required careful, measured restraint.

UNITY OF COMMAND (EFFORT)

C-22. During OPERATION RESTORE HOPE, UNITAF successfully met unity of command challenges through three innovations. First, they created a civil-military operations center (CMOC) to facilitate unity of effort between NGOs and military forces. Second, UNITAF divided the country into nine humanitarian relief sectors centered on critical urban areas that facilitated both relief distribution and military areas of responsibility. Third, to establish a reasonable span of control, nations that provided less than platoon-sized contingents were placed under the control of the Army, Marine Corps, and Air Force components.

C-23. On the other hand, during OPERATION CONTINUED HOPE, UNOSOM II command and control relationships made unity of command (effort) nearly impossible. The logistic components of USFORSOM were under UN operational control, while the QRF remained under CENTCOM's combatant command—as was TF Ranger. However, the CENTCOM commander was not in theater. He was not actively involved in planning TF Ranger's missions or in coordinating and integrating them with his other subordinate commands. It was left to TF Ranger to coordinate with the QRF *as needed*. Even in TF Ranger, there were dual chains of command between SOF operators and the Rangers. This underscores the need for close coordination and careful integration of SOF and conventional forces (see Chapter 4). It also emphasizes overall unity of command (or effort when command is not possible) among all forces operating in a single urban environment.

C-24. Following TF Ranger's 3 October mission, the command structure during OPERATION CONTINUED HOPE was further complicated with the new JTF-Somalia. This force was designed to protect US forces during the withdrawal from Somalia. JTF-Somalia came under the operational control of CENTCOM, but fell under the tactical control of USFORSOM. Neither the JTF nor USFORSOM controlled the naval forces that remained under CENTCOM's operational control. However unity of effort (force protection and a rapid, orderly withdrawal) galvanized the command and fostered close coordination and cooperation among the semiautonomous units.

MEASURED RESTRAINT

C-25. During OPERATIONS PROVIDE RELIEF and RESTORE HOPE, US forces dominated within their supporting roles. Their perseverance, adaptability, impartiality, and restraint allowed them to provide a stable, secure environment. Hence, relief organizations could provide the food and medical care necessary to reduce disease, malnourishment, and the overall mortality rate. However, during OPERATION CONTINUED HOPE, US operations became increasingly aggressive under the UN mandate. Peace enforcement also requires restraint and impartiality to successfully dominate and achieve political objectives. The increased use of force resulted in increased civilian casualties, which in turn reduced the Somalis' perception of US legitimacy. As a result, most moderate Somalis began to side with the Aideed and his supporters. Many Somalis felt that it was fine to intervene in the country to feed the starving and even help establish a peaceful government, but not to purposefully target specific Somali leaders as criminals.

TRANSITION

C-26. Across the spectrum of conflict, Army forces must be able to execute the full range of operations not only sequentially but, as in the case of operations in Somalia, simultaneously. OPERATION PROVIDE RELIEF began primarily as foreign humanitarian assistance (a support operation) and progressed to include peacekeeping (a stability operation), defensive operations to protect UN forces and relief supplies, and minimum offensive operations. As operations transitioned to OPERATION RESTORE HOPE, it became apparent that while foreign humanitarian assistance was still the principal operation, other operations were necessary. Peacekeeping, show of force, arms control, offensive, and defensive operations grew more necessary to establish a secure environment for uninterrupted relief operations. In the final phase of US involvement during OPERATION CONTINUED HOPE, major changes to political objectives caused a transition to peace enforcement with an increase in the use of force, offensively and defensively, to create a peaceful environment and conduct nation building.

SUMMARY

C-27. OPERATIONS PROVIDE RELIEF and RESTORE HOPE were unquestionably successes. Conversely, during OPERATION CONTINUED HOPE, the 3-4 October battle of Mogadishu (also known as the "Battle of the Black Sea") was a tactical success leading to an operational failure. TF Ranger succeeded in capturing 24 suspected Aideed supporters to include two of his key lieutenants. Arguably, given the appropriate response at the strategic level, it had the potential to be an operational success. After accompanying Ambassador Oakley to a meeting with Aideed soon after the battle, MG Zinni described Aideed as visibly shaken by the encounter. MG Zinni believed Aideed and his subordinate leadership were tired of the fighting and prepared to negotiate. Unfortunately, the US strategic leadership failed to conduct the shaping actions necessary to inform and convince the American public (and its elected members of Congress) of the necessity of employing American forces to capture Aideed. The president was left with little recourse after the battle of Mogadishu but to avoid further military confrontation.

C-28. Despite this strategic failing, the operational commanders might have avoided the casualties, and any subsequent public and Congressional backlash, had they better communicated among themselves and worked with unity of effort. Recognizing the separate US and UN chains of command, the UN Special Representative, along with the CENTCOM, USFORSOM, and TF Ranger commanders, should have established the command and control architecture needed. This architecture would have integrated planning and execution for each urban operation conducted. These commanders failed to "operationalize" their plan. They did not properly link US strategic objectives and concerns to the tactical plan. The TF Ranger mission was a direct operational attempt to obtain a strategic objective in a single tactical action. Yet, they failed to assess the lack of strategic groundwork, the threat's intent and capabilities, and the overall impact of the urban environment, to include the terrain and society, on the operation. Such an assessment may not have led to such a high-risk course of action and instead to one that de-emphasized military operations and emphasized a political solution that adequately considered the clans' influence.

Appendix D

Joint and Multinational Urban Operations

[Joint force commanders] synchronize the actions of air, land, sea, space, and special operations forces to achieve strategic and operational objectives through integrated, joint campaigns and major operations. The goal is to increase the total effectiveness of the joint force, not necessarily to involve all forces or to involve all forces equally.

JP 3-0

As pointed out earlier, Army forces, division size and larger, will likely be required to conduct operations in and around large urban areas in support of a joint force commander (JFC). The complexity of many urban environments, particularly those accessible from the sea, requires unique leveraging and integration of all the capabilities of US military forces to successfully conduct the operation. This appendix discusses many of these capabilities; JP 3-06 details joint urban operations.

PURPOSE

D-1. In some situations, a major urban operation is required in an inland area where only Army forces are operating. Army commanders determine if the unique requirements of the urban environment require forming a joint task force (JTF) or, if not, request support by joint capabilities from the higher joint headquarters. Sometimes the nature of the operation is straightforward enough or the urban operation is on a small enough scale that conventional intraservice support relationships are sufficient to meet the mission requirements.

D-2. Most major urban operations (UO), however, require the close cooperation and application of joint service capabilities. A JTF may be designated to closely synchronize the efforts of all services and functions in an urban area designated as a joint operations area (JOA). If a large urban area falls in the context of an even larger ground force area of operations, a JTF dedicated to the urban operation may not be appropriate. These

CONTENTS

situations still require joint capabilities. In such cases, the responsible JFC designates support relations between major land units and joint functional commands. The major land units can consist of Army forces, Marine Corps forces, or joint forces land component command. The joint functional commands can consist of the joint special operations task force (JSOTF), joint psychological operations (PSYOP) task force, or joint civil-military operations task force.

D-3. This appendix describes the roles of other services and joint combatant commands in UO. It provides an understanding that enables Army commanders to recommend when to form a JTF or to request support from the JFC. It also provides information so commanders can better coordinate their efforts with those of the JFC and the commanders of other services or components conducting UO. Lastly, this appendix describes some considerations when conducting UO with multinational forces.

SERVICE URBAN CAPABILITIES

D-4. Army forces conducting UO rely on other services and functional joint commands for specialized support in the urban environment. These capabilities are requested from and provided through the commanding JFC. Army forces request the assets and capabilities described in this annex through their higher headquarters to the joint command. The JFC determines if the assets will be made available, the appropriate command relationship, and the duration of the support. Army forces prepare to coordinate planning and execution with other services and to exchange liaison officers. These capabilities can greatly increase the Army's ability to assess, shape, dominate, and transition within the context of UO.

AIR FORCE

D-5. Air Force support is an important aspect of the Army force concept for urban operations. Air Force elements have a role to play in UO across the range of Army operations.

D-6. Air Force intelligence, surveillance, and reconnaissance (ISR) systems contribute significantly to assessing the urban area. These ISR systems include the E-8 Joint Surveillance, Target Attack Radar System (JSTARS) (see Figure D-1), U2S, RC-135 Rivet Joint, or RQ-4A Global Hawk unmanned aerial vehicle. Air Force ISR systems can provide vital data to help assess threat intentions, threat dispositions, and an understanding of the civilian population. These systems also can downlink raw information in real-time to Army intelligence processing and display systems, such as the common ground station or division tactical exploitation system.

Figure D-1. USAF E-8 JSTARS Platform

D-7. Air interdiction (AI) can be a vital component of shaping the urban battlespace. Often, AI of the avenues of approach into the urban area isolates the threat by diverting, disrupting, delaying, or destroying threat forces before they can be used effectively against Army forces. AI is especially effective in major theater war (MTW) circumstances where restrictions on airpower are limited and the threat is likely to be a conventionally equipped enemy. In 1991 during OPERATION DESERT STORM, AI helped prevent the Iraqi 5th Mechanized Division from reaching Khafji.

D-8. Precise air-delivered fires can positively and directly affect the conduct of Army close combat actions in the urban area. Special munitions designed to penetrate hardened bunkers can provide unique support to land forces executing UO. Problems associated with dense smoke and dust clouds hanging over the urban area and laser scatter may restrict the use of special, heavy, laser-guided bombs. If the launching aircraft can achieve a successful laser designation and lock-on, these weapons have devastating effects, penetrating deep into reinforced concrete before exploding with great force. If launched without a lock-on, or if the laser spot is lost, these weapons are unpredictable and can travel long distances before they impact.

D-9. General-purpose bombs from 500 to 2,000 pounds can also be used; however, they are only moderately effective against enemy located in large buildings. High-dive angle bomb runs may improve accuracy and penetration but will increase the aircraft's exposure to antiaircraft weapons. Low-dive angle bomb runs using high-drag (retarded) bombs may be appropriate to place bombs into upper stories but penetration is poor. On the other hand, aerial bombs can pass completely through light-clad buildings and explode on the outside with unwanted effects.

D-10. In addition to shaping and dominating the UO through firepower, commanders can use Air Force capabilities to improve and augment the urban transportation and distribution infrastructure. Air Force units can repair or improve airfields, revitalize civil aviation maintenance facilities, manage air-delivered cargo, and control civil and military air traffic. These latter capabilities particularly enhance urban stability operations and support operations. These capabilities may even be decisive. Air-delivered cargo and air traffic management, for example, were the decisive factors in US forces' successful resistance of the Soviet blockade of Berlin in 1948. In OPERATION RESTORE HOPE, from December 1992 to May 1993, Air Force operations in Mogadishu were critical to airlifting and staging supplies and forces. The Air Force determined the ultimate success of the humanitarian assistance operation (see the Somalia vignette in Appendix C).

D-11. In unique situations, such as the Berlin Blockade and OPERATION RESTORE HOPE, exercising Air Force urban capabilities may be the decisive action of the operation. Air Force capabilities will play a shaping role, sustaining role, or both in joint urban operations because of the requirement to occupy terrain and interface with the population.

MARINE CORPS

D-12. The Marine Corps can assault across water obstacles into a defended urban environment. This capability is an invaluable tactical and operational tool. The mere threat of this capability can divert many enemy forces from other avenues of approach and obscure the true nature of an attack. The impact of the threat of amphibious assault was vividly demonstrated during

OPERATION DESERT STORM where embarked Marine Forces diverted several Iraqi divisions to defensive positions along the coast and near Kuwait City.

D-13. The presence of Marine amphibious equipment, apart from Marine infantry, provides Army forces unique capabilities. In UO, the amphibious operation is often not an assault from the sea, but rather an assault river crossing. In 1950, the 7th US Infantry Division used amphibious tractor support from the 1st Marine Division to conduct an assault river crossing of the Han River into downtown Seoul.

D-14. The worldwide deployment of Marine air-ground task forces (MAGTFs) enables a short notice response into any urban areas accessible from the sea. Typically, a deployed MAGTF is a Marine expeditionary unit (special operations capable) (MEU[SOC]). The MEU(SOC) can perform forcible entry operations, seize lodgments, and may execute these tasks anticipating reinforcement by Army or joint forces. They are also well positioned and equipped to rapidly reinforce Army forces already deployed in theater. The special-operations-capable training that these units accomplish before deploying includes urban warfare training and contributes to their value in UO. The MEU(SOC) is relatively small (its core unit is a Marine infantry battalion), is forward deployed, and has a wide spectrum of organic capabilities. It is an important asset in crisis stability operations and support operations.

D-15. The Marine expeditionary brigade is the MAGTF between a MEU and a Marine expeditionary force (the largest MAGTF). All MAGTFs include an aviation combat element, which is particularly skilled at providing accurate and timely close air support. The responsiveness and accuracy of Marine close air support aptly suits it to UO where the lethality of combat and the close range of engagements demand accurate and responsive fires (see FM 3-31.1).

D-16. Marine forces that conduct UO work well in littoral urban areas because of their unique relationship with naval forces and thus their capability to closely integrate land and sea operations. A supporting arms liaison team may be attached to Army forces at battalion level to provide ship-to-shore communications and coordination for naval gunfire support.

NAVY

D-17. Many major urban areas are accessible from the sea. Army commanders understand how sea power can influence and support UO. The Navy brings several major capabilities to UO. These include naval gunfire support, naval air support, and port and coastal security.

D-18. Naval surface fire support (which includes naval gunfire support) particularly applies to forcible entry operations in littoral urban areas. It provides an initial indirect fire support capability until Army forces land ashore. However, naval gunfire support lacks a precision munitions capability. In defensive operations, naval gunfire support can also add major fires, especially during retrograde operations through an urban area. Naval gunfire support of the 3rd Infantry Division was the primary fire support for the last two days of the X Corps retrograde operation out of the port of Hungnam, North Korea, in December 1950.

Figure D-2. USN MK45 Lightweight Gun System

D-19. Destroyers and cruisers, which mount the 127mm MK45 lightweight gun system (see Figure D-2), usually provide naval gunfire support. This gun system can provide a rate of fire of 16-20 rounds per minute per gun to a range of approximately 23 kilometers. Naval air support, based on the highly mobile aircraft carrier battle group, can provide highly responsive full-spectrum aviation support to UO in most major urban areas. This alleviates the need for fixed-wing aviation bases.

D-20. Due to its flat trajectory, terrain masking affects naval gunfire more than field artillery. Naval gunfire also results in large range probable errors (the dispersion pattern of the naval gun is roughly elliptical with the long axis in the direction of fire). Hence, coverage of targets such as roads and airfields is most effective when the gun-target line (GTL) coincides with the long axis of the target. Very close supporting fire can be delivered when the GTL is parallel to the front line of troops. Oppositely, a GTL perpendicular to the front trace can endanger friendly forces. Within the limits of hydrographic conditions, the ship can maneuver to achieve a better GTL, but ship movement also makes it difficult to adjust fire. Overall, naval and air threats, bad weather, and large range probable errors make naval gunfire difficult and can cause cancellation of supporting fires.

D-21. Water terminals located in urban areas are usually the debarkation points for the bulk of Army forces as well as a joint force theater logistics requirement. Army forces are responsible for water terminal operations while naval capabilities protect these strategic and operationally vital facilities.

D-22. Naval coastal warfare (NCW) is the responsibility of the JFC and is often exercised through the Navy component commander. He may assign a naval coastal warfare commander for an appropriate geographic area. NCW includes coastal sea control, port security, and harbor defense. While coastal sea control is conducted in the environment of the open seas, port security and harbor security include the urban environment. Port security is the safeguarding of vessels, harbors, ports, waterfront facilities, and cargo from internal threats. It includes destruction, loss, or injury from sabotage or other subversive acts; accidents; thefts; or other causes of similar nature. The Navy's role in protecting essential urban infrastructure is often key. This task is also important when executing stability operations or support operations if the threats against urban infrastructure will likely be unconventional.

D-23. Harbor defense protects harbor approaches, harbors, anchorages, and ports from external threats. Harbor defense focuses on the conventional defense of port infrastructure. It is a task appropriate in a MTW scenario and often includes port security as a subtask. The JFC executes NCW (focused on harbor defense and port security) using combined Navy capabilities including surface warfare, aviation, and naval special operations. See also discussions in this appendix on transportation command and special operations command.

COAST GUARD

D-24. The Coast Guard, like the Navy, can significantly influence the conduct of UO when the urban area is accessible from the sea. The US Coast Guard (USCG) is the federal authority for port security and harbor defense of domestic facilities. When directed by the President, the USCG can augment the Navy in operations overseas. Historically, the entire USCG was under Navy control during both World Wars I and II. USCG elements deployed overseas and operated under Navy control during the Vietnam War and during OPERATIONS DESERT SHIELD and DESERT STORM.

D-25. The USCG uses surface warfare systems and aerial reconnaissance systems to conduct its missions. Its air systems are unarmed, whereas its surface systems are armed for self-defense and law enforcement operations. The USCG is experienced and adept at supporting other agencies, local governments, and law enforcement. Its capabilities can best support Army UO in stability operations or support operations or the stability and support aspects of an operation. The USCG works effectively against an unconventional threat and threats with on-water capability.

URBAN FUNCTIONAL COMBATANT COMMAND CAPABILITIES

D-26. Three commands provide urban functional combatant command capabilities. Transportation Command works for the Department of Defense. Space Command integrates several agencies. Special Operations Command uses special operations forces.

TRANSPORTATION COMMAND

D-27. US Transportation Command (USTRANSCOM) provides strategic air, land, and sea transportation for the Department of Defense to deploy, employ, sustain, and redeploy US military forces worldwide. USTRANSCOM provides global transportation management, using an integrated transportation system across the spectrum of operations through its transportation component commands (TCCs). The TCC consists of Air Mobility Command (AMC), Military Sealift Command (MSC), and Military Traffic Management Command (MTMC).

D-28. During urban operations, USTRANSCOM, through its TCC, can provide common-user terminal services in support of strategic transportation movements to a theater of operations. AMC provides common-user airlift, air refueling, and aeromedical evacuation services. It is the worldwide aerial port manager and, where designated, the operator of common-user aerial ports of embarkation and aerial ports of debarkation. MSC provides common-user sealift services between seaports of embarkation (SPOEs) and seaports of debarkation (SPODs). MTMC provides common-user ocean terminal service and, where designated, serves as the single port manager at SPOEs and SPODs. The urban transport system, in the form of railheads, ports, and airfields, is integral to many urban operations across the spectrum of Army operations and often the objective of Army UO. Army forces planning, preparing, and executing urban operations engage the supported geographic commander of a combatant command or his component commands in all aspects of conducting UO when the objective is transport related.

D-29. In offensive operations, commanders use transportation expertise to identify the urban transportation infrastructure (both the terrain and social [human] aspects) that is secured or that can affect current and future operations. This analysis also includes second- and third-order support systems. USTRANSCOM is consulted regarding the degree of acceptable damage that the system can sustain and still meet mission requirements. During execution, USTRANSCOM units may integrate into the operation so they can begin operating the transportation systems as early as possible. During the Inchon landing of September 1950, Army forces had begun rail operations on D+1 and port operations under way by D+3 of the forcible entry.

D-30. In defensive or stability operations, USTRANSCOM units safeguard and prevent disruption of the transport system by the conduct of defensive operations or stability tasks. USTRANSCOM advises Army commanders of the impact of defensive tactics, techniques, and procedures (TTP) on USTRANSCOM operations as well as security requirements for USTRANSCOM facilities. The JFC provides guidance to deconflict any issues that may arise from these potentially divergent missions and tasks.

D-31. Support UO, particularly in foreign humanitarian relief operations, often rely on USTRANSCOM to establish terminal operations and the tempo of transportation flow into a theater. The Army forces' planning, preparation, and execution of tasks, such as security and distribution actions in the area, support the geographic combatant commander's validated movement plan.

SPACE COMMAND

D-32. The United States Space Command (USSPACECOM) is the combatant command that integrates and synchronizes space capabilities to ensure their most effective use. USSPACECOM operates assigned space forces through its service component commands: US Army Space Command, Naval Space Command, and 14th Air Force (the Air Force space component).

D-33. Space systems offer global coverage and potential for real-time and near real-time support to military operations. Space systems are unconstrained by political boundaries. Commanders can use space systems during peacetime or times of crisis to monitor an urban area before inserting friendly forces. Space forces support Army UO through space force enhancement operations that provide products and services to multiply Army force effectiveness. Space support teams may assist both Army and JTF commanders to leverage system capabilities to facilitate planning and executing space operations.

D-34. Space systems enhance operations and assist commanders in overcoming some of the physical challenges of the urban environment. These enhancement operations include ISR; positioning and navigation; environmental monitoring; and communications. Space-based imagery and sensors are important ISR capabilities that contribute to situational understanding throughout the depth, breadth, and height of the urban area. ISR systems can provide route and target information for mission planning, locate presurveyed missile launch sites, detect camouflage, assess threat operations and movements, and warn of hostile acts and reconnaissance. Positioning and navigation systems also assist situational understanding through links to digital information systems (INFOSYS) while assisting tactical navigation in ambiguous terrain found in some urban areas. These enhancements can include precise location and position information for urban fires, ingress and egress routes, and rendezvous coordination. Environmental monitoring systems can provide weather and ionospheric information needed to assess weapon selection, air routes, ground and water trafficability, and communications. Communications systems provide secure, survivable links between elements of Army and joint forces to disseminate plans, orders, and warnings. These systems may form a critical link in the INFOSYS that transmit data to assessment centers and intelligence to key decisionmakers. However, some of the same environmental influences and degradation as ground-based systems may affect space-based systems.

SPECIAL OPERATIONS COMMAND

D-35. US Special Operations Command (USSOCOM) exercises combatant command of all active and reserve special operations forces stationed in the United States. USSOCOM also provides trained and combat-ready special operations forces (SOF) to the geographic combatant commanders and, when directed by the President or the secretary of defense, command designated special operations.

D-36. SOF provide commanders with capabilities critical to success in the urban environment. The density of this environment in both space and time requires the careful integration of SOF and conventional forces. Army forces conducting UO have a clear and unambiguous command and control (C2) relationship with the SOF in the urban area to ensure coordination, massing of effects, and unity of effort.

D-37. Each service has unique special operations (SO) capabilities. For example, SOF can identify and seize or destroy key terrain or infrastructure in denied areas; secure or capture key personnel; counter urban insurgencies; and conduct unconventional warfare in enemy-held urban areas. SOF can also emplace sensors, provide clandestine intelligence collection, and provide target acquisition information in the highly restrictive terrain of the urban environment. In multinational UO, they can provide coalition support teams with trained, culturally aware, language proficient, military liaison personnel with organic communications connectivity. Army leaders understand the SO capabilities available to ensure that they request the right support for Army UO and to ensure unity of effort within the urban JOA.

Army SOF

D-38. The Army provides five types of SOF units to USSOCOM. Two of these, PSYOP and civil affairs, are discussed in Chapter 4.

D-39. **Ranger Forces.** The US Army 75th Ranger Regiment can rapidly deploy light infantry forces from company through regimental size. Rangers specialize in direct action (DA) SO missions and focus on airfield seizure and raids. Typically, Ranger units turn over their objectives to conventional units upon mission completion. The Ranger DA capability especially applies to UO because of the many critical infrastructures that often prompt Army forces to engage in UO. Ranger capabilities are ideal for seizing critical facilities to preempt their defense or destruction. The size and combat power of Ranger units permit execution of offensive and defensive operations against enemy conventional units for periods of limited duration; austere combat service support capabilities limit the Ranger regiment's ability to sustain combat action without extensive augmentation.

D-40. **Special Forces.** US Army Special Forces Command (Airborne) trains and prepares Army Special Forces (SF) to deploy and execute operational requirements for geographic combatant commanders. SF units are small and capable of extended operations in remote and hostile locations. SF units execute seven basic missions: foreign internal defense, unconventional warfare, counterproliferation, special reconnaissance (SR), DA, combatting terrorism, and information operations. Important collateral activities consist of coalition support, combat search and rescue, counterdrug activities, humanitarian demining activities, humanitarian assistance, security assistance, and special activities.

D-41. SR, DA, and coalition support are particularly important in UO. SF units may physically penetrate an urban area to conduct SR to determine threat strengths, dispositions, and intentions. In some situations, SF units, due to their language and cultural training, can gather accurate information regarding the disposition and attitudes of the population. SF special reconnaissance can also determine or verify the functional status of urban infrastructure as well as conduct target acquisition, area assessment, and post-strike reconnaissance.

D-42. SF teams can execute terminal guidance or control operations for Army and joint precision fires. Although these teams possess limited organic combat power for their short-duration, DA missions, they can bring significant effects to bear against high-payoff targets. Yet, SF units have no capability to conduct a sustained defense of such targets.

D-43. SF units can advise, train, and assist urban indigenous movements already in existence to conduct unconventional warfare and possibly accompany these groups into combat. The upsurge in urban insurgency and terrorism has caused worldwide concern since it is not confined to developing countries. Present day dissident groups are well aware of this situation. These groups realize that to be successful, they must center the insurgent activities on the major cities or political center of their countries.

D-44. **Special Operations Aviation Forces.** Army special operations aviation provided by the 160th Special Operations Aviation Regiment operates primarily to support SO missions. They execute insertion, extraction, and resupply missions to support SO. Similar to conventional aviation, this capability may be more vulnerable to concealed air defenses when operating over hostile or unsecured urban terrain than in many other environments.

Navy SOF

D-45. **Sea-Air-Land Teams.** Navy sea-air-land teams (SEALs) specialize in water approaches to targets. They operate in small, squad-size teams and have many of the same capabilities as Army Special Forces (see above). Navy SEALs do not typically have the cultural and language training of Army Special Forces. Their capability to insert from sea gives them a unique ability to penetrate into urban areas that are accessible from the sea.

D-46. **Special Boat Units.** Special boat units (SBUs) employ, operate, and maintain various surface combatant craft (see Figure D-3) to conduct and support naval and joint special operations, riverine warfare, and coastal patrol and interdiction. The SBU can infiltrate and exfiltrate forces; provide small-caliber gunfire support; and conduct coastal patrol, surveillance, harassment, and interdiction of maritime lines of communications. These units are ideal in UO that include ports and rivers. They can assist in port security, conduct river patrols, and participate in harbor defense. They are well suited for preventing sea infiltration by unconventional threats.

Figure D-3. USN MK V Special Operations Craft

Air Force SOF

D-47. The Air Force has two primary SOF elements: special operations air units and special tactics units.

D-48. **Special Air Wings.** Air Force special operations aviation elements operate both fixed-wing and rotary-wing aircraft (C-130 variants, MH-60, and MH-53). The air elements perform various missions including rotary-wing refueling, single aircraft penetration, close air support of SOF ground elements, PSYOP support, aerial insertion of ground SOF, electronic warfare, and aerial C2 support.

D-49. The AC-130 U and H model aircraft (see Figure D-4 on page D-10) are designed specifically to provide close air support to ground SOF. They are armed with one 40mm autocannon and one 105mm howitzer. The AC-130 U is also armed with one 25mm autocannon. This aircraft has night capability and is extremely accurate. Its fires are responsive and can be decisive in close urban combat. Its cannon and howitzer are accurate enough to concentrate fire onto a single spot to create a rooftop breach that allows fire to be directed deep into the building. The AC-130 is very vulnerable to air defense systems, which friendly forces must suppress or destroy to effectively use this system.

(During OPERATION DESERT STORM in 1991, an AC-130 was shot down over Kuwait.)

Figure D-4. USAF AC-130 Gunship

D-50. **Special Tactics Forces.** Air Force special tactics forces consist of special tactics teams (STTs) and special operations weather teams (SOWTs). Combat controllers and pararescue personnel comprise STTs. STTs are specially tailored to meet mission criteria and may vary from a small three-man team to a larger twenty-five-man element. STTs support the UO Army commanders by—

D-51.

D-52.

may be assigned tasks that require close relations with civilians. Army forces, in this situation, may be assigned tasks that are more remote from the population. National forces that have a national history of animosity to the civil population (or that sympathize with antagonists within the civilian population) are not used in tasks requiring diplomacy and close cooperation with the civilians or government.

D-53. Army forces are responsible to understand the military capabilities of national forces with which they work. Some national forces, as part of their normal capabilities, are adept at police functions that enable them to operate with little training in a law enforcement role. Other national forces specialize in small-unit, light infantry patrolling. These forces may be ideal in a stability operation. In contrast, a national force comprised of conscripts and trained primarily in conventional warfare techniques may best work as a reinforcing force or may require extensive training before mission execution in an urban environment or a stability situation. Army commanders also consider the type of weapon systems with which participating multinational forces are equipped; they may be more or less effective in an urban environment. For example, some countries may still possess the Vulcan antiaircraft gun system (or similar weapon) that can be very useful in urban offensive and defensive operations.

D-54. Logistically, agreement among the multinational forces should include support consolidation whenever possible. Creating a multinational logistics office works best when accomplished early. This office can coordinate local contracts as well as already agreed upon host-nation support. This coordination among participating nations will reduce the competition for local assets that could otherwise have detrimental effects on one or more participating nations.

Source Notes

These are the sources quoted or paraphrased in this publication. They are listed by page number. Boldface indicates the titles of historical vignettes.

Chapter 1—Urban Outlook

1-1 "The ambiguous nature of the operational environment . . . .": FM 1, *The Army,* 14 June 2001: 12.

1-4 **Rome: A Microcosm of Urban Warfare**: R. Ernest Dupuy and Trevor N. Dupuy, *The Encyclopedia of Military History: From 3500 B.C. to the Present* (San Francisco: Harper & Row, 1977).

Chapter 2—Urban Environment

2-1 "From a planning perspective, commanders view cities . . . .": FM 3-0, *Operations,* 14 June 2001: 6-19.

2-17 "Me and Somalia against the world, ...": cited in Andrew S. Natsios, "Food Through Force: Humanitarian Intervention and U.S. Policy," *The Washington Quarterly* (winter 1994): 136.

2-21 "Whoever coined the phrase 'The Theatre of Operations' . . . ": General Sir Rupert Smith, "Wars in Our Time–A Survey of Recent and Continuing Conflicts," *World Defence Systems* (July 2001): 121.

2-25 **Seeing the Urban Area and Its Parts**: Russell W. Glenn, "Urban Combat is Complex." *Proceedings (*February 2002): 65.

Chapter 3—Urban Threat

3-1 ". . . [T]he United States could be forced to intervene . . . .": *Quadrennial Defense Review Report,* 30 September 2001: 6.

3-4 **Tempo**: 26th Infantry Regimental Association, *Aachen. Military Operations in Urban Terrain* (Lititz, PA: 26th Infantry Regimental Association, 1999); Irving Werstein, *The Battle of Aachen* (New York: Thomas Y. Crowell Company, 1962); Charles Whiting, *Bloody Aachen* (New York: Stein and Day, 1976).

3-7 "Chechen fighters sometimes disguised . . . .": Olga Oliker, *Russia's Chechen Wars 1994–2000: Lessons From Urban Combat* (Santa Monica, CA: RAND, 2001): 21.

3-8 **Identifying Soldiers from Civilians**: Olga Oliker, *Russia's Chechen Wars 1994–2000: Lessons From Urban Combat* (Santa Monica, CA: RAND, 2001).

3-9 **Information and the Media**: Eric Hammel, *Fire in the Streets: The Battle for Hue, Tet 1968* (New York: Dell Publishing, 1991).

3-13 **Cultural and Religious Instability**: Sid Heal, "Crowds, Mobs and Nonlethal Weapons," *Military Review* (March/April 2000): 45–50.

3-14 **Food and Water Shortages**: Leif R. Rosenberger, "The Strategic Importance of the World Food Supply," *Parameters* (spring 1997): 84–105.

3-16 **Urban Insurgencies**: Sid Heal, "Crowds, Mobs and Nonlethal Weapons," *Military Review* (March/April 2000): 45–50; Lawrence A. Yates, "Military Stability and Support Operations: Analogies, Patterns and Recurring Themes," *Military Review* (July/August 1997): 51–61.

3-17 **Crime and Criminal Organizations**: Eugene Linden, "The Exploding Cities of the Developing World," *Foreign Affairs* (January/February 1996): 52–65; Andrew S. Natsios, "Commander's Guidance: A Challenge of Complex Humanitarian Emergencies," *Parameters* (summer 1996): 50–66.

Chapter 4—Contemplating Urban Operations

4-1 "We based all our further calculations . . . .": S. M. Shtemenko, *The Soviet General Staff at War, 1941-1945: Book One* (Moscow: Progress Publishers, 1981): 317–318.

4-23 "Four hostile newspapers . . . .": Napoleon Bonaparte, Justin Wintle (ed.), *The Dictionary of War Quotations* (New York: The Free Press, 1989): 73.

Chapter 5—Foundations for Urban Operations

5-1 "Utilities such as electricity . . . .": Richard Connaughton, John Pimlott, and Duncan Anderson, *The Battle for Manila: The Most Devastating Untold Story of World War II* (Novata, CA: Presidio Press, 1995): 103.

5-9 **Applying the Urban Operational Framework: Panama – December 1989**: Ronald H. Cole, *Operation Just Cause* (Washington DC: Joint History Office, 1995); Thomas Donnelly, Margaret Roth, and Caleb Baker, *Operation Just Cause: The Storming of Panama* (New York: Lexington Books, 1991); John T. Fishel, *The Fog of Peace: Planning and Executing the Restoration of Panama* (Carlisle, PA: Strategic Studies Institute, 1992); Malcolm McConnell, *Just Cause: The Real Story of America's High-Tech Invasion of Panama* (New York: St. Martin's Press, 1991); John Embry Parkerson, Jr., "United States Compliance with Humanitarian Law Respecting Civilians During Operation Just Cause," *Military Law Review* (1991): 31–140; Jennifer Morrison Taw, *Operation Just Cause: Lessons for Operations Other Than War* (Santa Monica, CA: RAND, 1996).

5-26 "Fighting in a city is much . . . .": Vasili I. Chuikov, *The Battle for Stalingrad* (New York: Holt, Rinehart, and Winston, 1964): as cited by JP 3-06, *Doctrine for Urban Operations*, 16 September 2002: 1-8.

5-28 "I heard small-arms fire and . . . .": Mark A. B. Hollis, "Platoon Under Fire," *Infantry* (January–April 1998): 29–30.

5-31 **Example of Simple Communications Innovation: Israel's Six-Day War – 1967**: Trevor N. Dupuy and Paul Martell, *Flawed Victory: The Arab-Israeli Conflict and the 1982 War in Lebanon* (Fairfax, VA: Hero Books, 1986); Richard A. Gabriel, *Operation Peace for Galilee: The Israeli-PLO War in Lebanon* (New York: Hill and Wang, 1984).

Chapter 6—Urban Offensive Operations

6-1 " . . . Capture Suez City . . . .": Abraham Adan, *On the Banks of the Suez* (Novato, CA: Presidio Press, 1980): 409.

6-3 **The Operational Context of Urban Operations: Brittany Ports – August to September 1944**: Kent H. Butts, "The Strategic Importance of Water," *Parameters* (spring 1997): 65–83; Paul Carell, *Stalingrad: The Defeat of the German 6th Army* (Atglen, PA: Schieffer Publishing, 1993); R. Ernest Dupuy and Trevor N. Dupuy, *The Encyclopedia of Military History: From 3500 B.C. to the Present* (San Francisco: Harper & Row, 1977); Ralph Peters, "Our Soldiers, Their Cities," *Parameters* (spring 1996): 43–50; John J. Peterson, *Into the Cauldron* (Clinton, MD: Clavier House, 1973).

6-12 **Forms of Attack in the Urban Offense: Metz – 1944**: Anthony Kemp, *The Unknown Battle: Metz, 1944* (New York: Stein and Day, 1981); John Colby, *War From the Ground Up: The 90th Division in WWII* (Austin, TX: Nortex Press, 1991).

6-21 **Isolating the Urban Area: Hue, Vietnam – January to February 1968**: James R. Arnold, *Tet Offensive 1968: Turning Point in Vietnam* (London: Osprey, 1990); Eric Hammel, *Fire in the Streets: The Battle for Hue, Tet 1968* (New York: Dell Publishing, 1991).

6-25 "Often what seems to be . . . .": Russell W. Glenn, *An Attack on Duffer's Downtown* (Santa Monica, CA: RAND, 2001): 28.

6-27 **Creative Task Organization: Using Artillery in the Direct Fire Role**: 26th Infantry Regimental Association, *Aachen. Military Operations in Urban Terrai* (Lititz, PA: 26th Infantry Regimental Association, 1999); Ronald H. Cole, *Operation Just Cause* (Washington DC: Joint History Office, 1995); Thomas Donnelly, Margaret Roth, and Caleb Baker, *Operation Just Cause: The Storming of Panama* (New York: Lexington Books, 1991); David Eshel, *Mid-East Wars: The Lebanon War 1982* (Hod Hasharon, Israel: Eshel-Dramit, 1983); Richard A. Gabriel, *Operation Peace for Galilee: The Israeli-PLO War in Lebanon* (New York: Hill and Wang, 1984); Malcolm McConnell, *Just Cause: The Real Story of America's High-Tech Invasion of Panama* (New York: St. Martin's Press, 1991); Irving Werstein, *The Battle of Aachen* (New York: Thomas Y. Crowell Company, 1962); Charles Whiting, *Bloody Aachen* (New York: Stein and Day, 1976); Bruce Allen Watson, *Sieges: A Comparative Study* (Westport, CT: Praeger, 1993).

6-30 **Bold Operational Maneuver to Seize an Urban Area: Inchon and Seoul, Korea – September 1950**: Robert Debs Heinl, Jr., *Victory at High Tide: The Inchon-Seoul Campaign* (Annapolis, MD: Nautical & Aviation Publishing Company of America, 1979); Douglas MacArthur, *Reminiscences* (New York: McGraw-Hill, 1964); Shelby L. Stanton, *Ten Corps in Korea, 1950* (Novato, CA: Presidio Press, 1989).

Chapter 7—Urban Defensive Operations

7-1 Generally, a modern city magnifies . . . .": Christopher R. Gabel, "Military Operations on Urbanized Terrain: The 2d Battalion, 26th Infantry, at Aachen, October 1944," *Center for Army Lessons Learned: Urban Combat Operations* (Newsletter No. 99-16). Available: http://call.army.mil/products/newsltrs/99-16/chap1.htm.

7-6 **Urban Defense in a Major Operation: Stalingrad – August 1942 to January 1943**: Anthony Beevor, *Stalingrad* (New York: Penguin Books, 1998); Vasili I. Chuikov, *The Battle for Stalingrad* (New York: Holt, Rinehart, and Winston, 1964); William Craig, *Enemy at the Gates: The Battle for Stalingrad* (New York: Reader's Digest Press, 1973).

7-10 "What is the position . . . .": Michael Dewar, *War in the Streets: The Story of Urban Combat from Calais to Khafji* (New York: Sterling Publishers, 1992): 8.

7-13 **Defensive Combat Power: Suez – October 1973**: Abraham Adan, *On the Banks of the Suez* (Novato, CA: Presidio Press, 1980).

Chapter 8—Urban Stability Operations And Support Operations

8-1 "The Rangers were bound . . . .": Mark Bowden, *Black Hawk Down: A Story of Modern War* (New York: Atlantic Monthly Press, 1999): 18.

8-6 "In wars of intervention . . . .": Antoine-Henri Jomini, "The Art of War," in *Roots of Strategy: Book 2* (Harrisburg, PA: Stackpole Books, 1987): 441.

8-9 **Assessment of Security and Force Protection: Belfast, Northern Ireland**: Michael Dewar, *British Army in Northern Ireland* (New York: Sterling Publishers, 1985); Desmond Hamill, *Pig in the Middle: The Army in Northern Ireland, 1969-1984* (London: Methuen, 1985).

8-12 "If there is any lesson . . . .": Frederick Funston, "How the Army Worked to Save San Francisco," *Cosmopolitan Magazine* (July 1906). Available http://www.sfmuseum.org/1906/cosmo.html

8-13 **Support of and Coordination with Civilian Authorities: The 1992 Los Angeles Riots**: James D. Delk, *Fires & Furies: The LA Riots* (Palm Springs, CA: Etc. Publications, 1995); William W. Mendel, *Combat in Cities: The LA Riots and Operation Rio* (Fort Leavenworth, KS: Foreign Military Studies Office, 1996).

Chapter 9—Urban Combat Service Support

9-1 "Even supply is different . . . ." Ralph Peters, "Our Soldiers, Their Cities," *Parameters* (spring 1996): 48.

9-3 **Base Security: Tan Son Nhut, Vietnam – Tet 1968**: James R. Arnold, *Tet Offensive 1968: Turning Point in Vietnam* (London: Osprey, 1990); Keith William Nolan, *The Battle for Saigon: Tet 1968* (New York: Pocket Books, 1996).

9-4 "CSS planning accounts for . . . .": FM 3-0, *Operations*, 14 June 2001: 6-19.

9-8 "Greater friendly force density would appear . . . .": Russell W. Glenn, Steven Hartman, and Scott Gerwehr, *On the Shoulders of Atlas: Urban Comabt Service Support Operations* (Santa Monica, CA: RAND, pending publication in 2003): 3.

9-19 **Combat Stress: Chechnya – 1994 to 1996**: Timothy L. Thomas and Charles P. O'Hara, "Combat Stress in Chechnya: 'The Equal Opportunity Disorder'," *Army Medical Department Journal* (January–March 2000): 46–53.

9-22 "Urban operations will continue . . . ": FM 1, *The Army*, 14 June 2001: 35.

Appendix A—Siege of Beirut: An Illustration of the Fundamentals of Urban Operations

A-1 **Siege of Beirut: An Illustration of the Fundamentals of Urban Operations**: Trevor N. Dupuy and Paul Martell, *Flawed Victory: The Arab-Israeli Conflict and the 1982 War in Lebanon* (Fairfax, VA: Hero Books, 1986); David Eshel, *Mid-East Wars: The Lebanon War 1982* (Hod

Hasharon, Israel: Eshel-Dramit, 1983); Richard A. Gabriel, *Operation Peace for Galilee: The Israeli-PLO War in Lebanon* (New York: Hill and Wang, 1984).

A-1 "The IDF had neither the strategy . . . .": Richard A. Gabriel, *Operation Peace for Galilee: The Israeli-PLO War in Lebanon* (New York: Hill and Wang, 1984): 128.

Appendix B—Urban Intelligence Preparation of the Battlefield

B-1 "Maneuvers that are possible . . . .": *Infantry in Battle,* 2nd ed. (Richmond, VA: Garret & Massie, 1939; Reprint, Fort Leavenworth, KS: US Army Command and General Staff College, 1981): 69.

B-6 **Analysis of an Urban Area's Underlying Terrain: Mitrovica, Kosovo**: Joseph D. Celeski, "Joint Urban Operations: Special Forces in Urban Campaign Planning," *Special Warfare* (fall 2001): 2–13; Russell W. Glenn, "Urban Combat is Complex," *Proceedings* (February 2002): 62–65.

B-12 " . . . [T]he adversaries of freedom . . . .": John F. Kennedy, Message to Congress, 25 May 1961. Available http://www.cs.umb.edu/jfklibrary/j052561.htm

B-14 **Shifting Civilian Interests and Intent**: James D. Delk, *Fires & Furies: The LA Riots* (Palm Springs, CA: Etc. Publications, 1995).

Appendix C—Operations in Somalia: Applying the Urban Operational Framework to Support and Stability

C-1 **Operations in Somalia: Applying the Urban Operational Framework to Support and Stability**: Kenneth Allard, *Somalia Operations: Lessons Learned* (Washington, DC: National Defense University Press, 1995).

C-1 "It's impossible for an American mother . . . .": Major General Thomas M. Montgomery. Available http://www.pbs.org/wgbh/pages/frontline/shows/ambush/interviews/montgomery.html

Appendix D—Joint and Multinational Urban Operations

D-0 "[Joint force commanders] synchronize the actions . . . ." JP 3-0, *Doctrine for Joint Operations*, 10 September 2001: II-4.

Glossary

The glossary lists acronyms and terms with Army or joint definitions. Where Army and joint definitions are different, *(Army)* follows the term. Definitions for which FM 3-06 is the proponent manual (the authority) are marked with an asterisk (*). The proponent or amplifying manual for other terms is listed in parentheses after the definition.

AD anno Domini (in the year of the Lord)

AF Air Force

A.F.B. Air Force Base (graphics only)

agility the ability to move and adjust quickly and easily (FM 3-0)

AI air interdiction

AIDS acquired immune deficiency syndrome

air defense battlefield operating system the employment of all active measures designed to nullify or reduce the effectiveness of attack by hostile aircraft and missiles after they are airborne (FM 7-15)

AMC Air Mobility Command

AO area of operations

area defense a type of defensive operation that concentrates on denying enemy forces access to designated terrain for a specific time rather than destroying the enemy outright (FM 3-0)

area of interest that area of concern to the commander, including the area of influence, areas adjacent thereto, and extending into enemy territory to the objectives of current or planned operations. This area also includes areas occupied by enemy forces that could jeopardize the accomplishment of the mission. (JP 3-0)

area of operations an operational area defined by the joint force commander for land and naval forces. Areas of operations do not typically encompass the entire operational area of the joint force commander, but should be large enough for component commanders to accomplish their missions and protect their forces. (JP 3-0)

ARFOR the senior Army headquarters and all Army forces assigned or attached to a combatant command, subordinate joint force command, joint functional command, or multinational command (FM 3-0)

ARVN Army of the Republic of Vietnam

assessment (Army) the continuous monitoring—throughout planning preparation, and execution—of the current situation and progress of an operation, and the evaluation of it against criteria of success to make decisions and adjustments (FM 3-0)

asymmetry dissimilarities in organization, equipment, doctrine, capabilities, and values between other armed forces (formally organized or not) and US forces. Engagements are symmetric if forces, technologies, and weapons are similar; they are asymmetric if forces, technologies, and weapons are different, or if a resort to terrorism and rejection of more conventional rules of engagement are the norm. (FM 3-0)

attack an offensive operation that destroys or defeats enemy forces, seizes and secures terrain, or both (FM 3-0)

attainability one of the eight characteristics of combat service support: generating the minimum essential supplies and services necessary to begin operations (FM 100-10)

attrition the reduction of the effectiveness of a force caused by loss of personnel and materiel (JP 1-02)

avenue of approach (Army) the air or ground route leading to an objective (or key terrain in its path) that an attacking force can use (FM 3-90)

axis of advance the general area through which the bulk of a unit's combat power must move (FM 3-90)

battle a set of related tactical engagements that last longer and involve larger forces than an engagement (FM 3-0)

battlefield operating systems the physical means that tactical commanders use to execute operations and accomplish missions assigned by superior tactical- and operational-level commanders. The seven battlefield operating systems are: the intelligence system, maneuver system, fire support system, air defense system, mobility/countermobility/survivability system, combat service support system, and command and control system. (FM 7-15)

battlefield organization the allocation of forces in the area of operations by purpose. It consists of three all-encompassing categories of operations: decisive, shaping, and sustaining. (FM 3-0)

battlespace the environment, factors, and conditions that must be understood to successfully apply combat power, protect the force, or complete the mission. This includes air, land, sea, space, and the included enemy and friendly forces; facilities; weather; terrain; the electromagnetic spectrum; and the information environment within the operational areas and areas of interest. (JP 3-0)

BC before Christ

BDAR battle damage assessment and repair

Bde brigade (graphics only)

Benelux an area of Europe comprised of three countries: Belgium, the Netherlands, and Luxembourg

Bns battalions (graphics only)

BOS battlefield operating systems

branch a contingency plan or course of action (an option built into the basic plan or course of action) for changing the mission, disposition, orientation, or direction of movement of the force to aid success of the current operation, based on anticipated events, opportunities, or disruptions caused by enemy actions. Army forces prepare branches to exploit success and opportunities, or to counter disruptions caused by enemy actions. (FM 3-0)

breakout an operation conducted by an encircled force to regain freedom of movement or contact with friendly units. It differs from other attacks only in that a simultaneous defense in other areas of the perimeter must be maintained. (FM 3-90)

buffer zone a defined area controlled by a peace operations force from which disputing or belligerent forces have been excluded. A buffer zone is formed to create an area of separation between disputing or belligerent forces and reduce the risk of renewed conflict. Also called area of separation in some United Nations operations. (JP 3-07.3)

bypass a tactical mission task in which the commander directs his unit to maneuver around an obstacle, position, or enemy force to maintain the momentum of the operation while deliberately avoiding combat with an enemy force (FM 3-90)

C2 command and control

CA civil affairs

campaign a series of related military operations aimed at accomplishing a strategic or operational objective within a given time and space (JP 3-0)

canalize (Army) a tactical mission task in which the commander restricts enemy movement to a narrow zone by exploiting terrain coupled with the use of obstacles, fires, or friendly maneuver (FM 3-90)

capability the ability to execute a specified course of action (a capability may or may not be accompanied by an intention) (JP 1-02)

casualty evacuation a term used by nonmedical units to refer to the movement of casualties aboard nonmedical vehicles or aircraft (FM 8-10-6)

CAV Cavalry (graphics only)

CBRNE chemical, biological, radiological, nuclear, and high-yield explosive

CCIR commander's critical information requirements

Cdrs commanders (graphics only)

CENTCOM US Central Command

center of gravity those characteristics, capabilities, or sources of power from which a military force derives its freedom of action, physical strength, or will to fight (JP 3-0)

CHS combat health support

CI counterintelligence

CINC commander in chief

civil affairs designated Active and Reserve component forces and units organized, trained, and equipped specifically to conduct civil affairs activities and to support civil-military operations (JP 3-57.1)

civil considerations how the attitudes and activities of the civilian leaders, populations, and organizations within an area of operations will influence the conduct of military operations (FM 6-0)

civil disturbances riots, acts of violence, insurrections, unlawful obstructions or assemblages, or other disorders prejudicial to public law and order. The term civil disturbance includes all domestic conditions requiring or likely to require the use of Federal Armed Forces pursuant to the provisions of Chapter 15 of Title 10, United States Code. (JP 3-07.3)

civil-military operations the activities of a commander that establish, maintain, influence, or exploit relations between military forces, governmental and nongovernmental civilian organizations and authorities, and the civilian populace in a friendly, neutral, or hostile operational area in order to facilitate military operations, to consolidate and achieve US objectives. Civil-military operations may include performance by military forces of activities and functions normally the responsibility of the local, regional, or national government. These activities may occur prior to, during, or subsequent to other military actions. They may also occur, if directed, in the absence of other military operations. Civil-military operations may be performed by designated civil affairs, by other military forces, or by a combination of civil affairs and other forces. (JP 3-57)

civil-military operations center an ad hoc organization, normally established by the geographic combatant commander or subordinate joint force commander, to assist in the coordination of activities of engaged military forces, and other United States Government agencies, nongovernmental organizations, and regional and international organizations. There is no established structure, and its size and composition are situation dependent. (FM 41-10)

clear (Army) 1. A tactical mission task that requires the commander to remove all enemy forces and eliminate organized resistance within an assigned area. (FM 3-90) 2. The total elimination or neutralization of an obstacle that is usually performed by follow-on engineers and is not done under fire. (FM 3-34.2)

close air support air action by fixed- and rotary-wing aircraft against hostile targets which are in close proximity to friendly forces and which require detailed integration of each air mission with the fire and movement of those forces (JP 3-09.3)

close combat combat carried out with direct-fire weapons, supported by indirect fires, air-delivered fires, and nonlethal engagement means. Close combat defeats or destroys enemy forces or seizes and retains ground. (FM 3-0)

CMO civil-military operations

CMOC civil-military operations center

CNA computer network attack

CND computer network defense

CNE computer network exploitation

CNO computer network operations

COA course of action

coalition an *ad hoc* arrangement between two or more nations for common action (JP 3-16)

COG center of gravity

collateral damage unintended and undesirable civilian personnel injuries or materiel damage adjacent to a target produced by the effects of friendly weapons (FM 6-30)

combatant command a unified or specified command with a broad continuing mission under a single commander established and so designated by the President, through the Secretary of Defense and with the advice and assistance of the Chairman of the Joint Chiefs of Staff. Combatant commands typically have geographic or functional responsibilities. (JP 0-2)

combat configured load a planned package of ammunition or other supplies that are transported as a single load to support a type unit or weapon system (FM 4-30.13)

combat power the total means of destructive and/or disruptive force which a military unit/formation can apply against the opponent at a given time (JP 3-0)

combat service support battlefield operating system the support and services provided to sustain forces during war and military operations other than war (FM 7-15)

combatting terrorism actions, including antiterrorism (defensive measures taken to reduce vulnerability to terrorist acts) and counterterrorism (offensive measures taken to prevent, deter, and respond to terrorism), taken to oppose terrorism throughout the entire threat spectrum (JP 3-07.2)

combined arms the synchronized or simultaneous application of several arms—such as infantry, armor, artillery, engineers, air defense, and aviation—to achieve an effect on the enemy that is greater than if each arm was used against the enemy separately or in sequence (FM 3-0)

combined arms team two or more arms mutually supporting one another, usually consisting of infantry, armor, cavalry, aviation, field artillery, air defense artillery, and engineers (FM 3-90)

command and control battlefield operating system all tasks associated with the exercise of authority and direction by a properly designated commander over assigned and available forces in the accomplishment of the mission (FM 7-15)

command and control system (Army) the arrangement of personnel, information management, procedures, and equipment and facilities essential to the commander to plan, prepare for, execute, and assess operations (FM 6-0)

commander's critical information requirements (Army) elements of information required by commanders that directly affect decision making and dictate the successful execution of military operations (FM 3-0)

commander's intent a clear, concise statement of what the force must do and the conditions the force must meet to succeed with respect to the enemy, terrain, and the desired end state (FM 3-0)

commander's visualization the process of developing a clear understanding of the current state with relation to the enemy and environment, envisioning a desired end state which represents mission accomplishment, and then subsequently visualizing the sequence of activity that moves the force from its current state to the end state (FM 6-0)

common operational picture an operational picture tailored to the user's requirements, based on common data and information shared by more than one command (FM 3-0)

computer network attack operations to disrupt, deny, degrade, or destroy information resident in computers and computer networks, or the computers and networks themselves (JP 3-13)

computer network defense defensive measures to protect and defend information, computers, and networks from disruption, denial, degradation, or destruction (JP 3-13)

concept of operations (Army) describes how commanders see the actions of subordinate units fitting together to accomplish the mission. As a minimum, the description includes the scheme of maneuver and concept of fires. It expands the commander's selected course of action and expresses how each element of the force will cooperate to accomplish the mission. (FM 3-0)

constraint restrictions placed on the command by a higher command to dictate an action or inaction, thus restricting the freedom of action the subordinate commander has for planning a mission by stating things that must or must not be done (FM 101-5)

contain (Army) a tactical mission task that requires the commander to stop, hold, or surround enemy forces or to cause them to center their activity on a given front and prevent them from withdrawing any part of their forces for use elsewhere (FM 3-90)

control (Army) 1. The regulation of forces and operating systems to accomplish the mission in accordance with the commander's intent. It includes collecting, processing, displaying, storing, and disseminating information for creating the common operational picture (COP) and using information, primarily by the staff, during planning, preparing for, and executing operations. (FM 6-0). 2. A tactical mission task that requires the commander to maintain physical influence over a specified area to prevent its use by an enemy or to create conditions necessary for successful friendly operations (FM 3-90)

controlled exchange the removal of serviceable parts, components, or assemblies from unserviceable, economically reparable equipment and their immediate reuse in restoring a like item of equipment to a combat operable or serviceable condition (FM 4-30.3)

control measures directives given graphically or orally by a commander to subordinate commands to assign responsibilities, coordinate fires and maneuver, and control combat operations. Each control measure can be portrayed graphically. In general, all control measures should be easily identifiable on the ground. Examples of control measures include boundaries, objectives, coordinating points, contact point, and direction of attack. (FM 101-5)

COP common operational picture

counterattack (Army) a form of attack by part or all of a defending force against an enemy attacking force, with the general objective of denying the enemy his goal in attacking (FM 3-0)

counterdeception efforts to negate, neutralize, diminish the effects of, or gain advantage from a foreign deception operation. Counterdeception does not include the intelligence function of identifying foreign deception operations. (JP 3-58)

counterdrug those active measures taken to detect, monitor, and counter the production, trafficking, and use of illegal drugs (JP 3-07.4)

counterinsurgency those military, paramilitary, political, economic, psychological, and civic actions taken by a government to defeat insurgency (JP 3-07)

counterintelligence information gathered and activities conducted to protect against espionage, other intelligence activities, sabotage, or assassinations conducted by or on behalf of foreign governments or elements thereof, foreign organizations, or foreign persons, or international terrorist activities (JP 2-01.2)

countermobility operations the construction of obstacles and emplacement of minefields to delay, disrupt, and destroy the enemy by reinforcement of the terrain. The primary purpose of countermobility operations is to slow or divert the enemy, to increase time for target acquisition, and to increase weapon effectiveness. (JP 3-34)

counterpropaganda activities that identify and counter adversary propaganda by exposing adversary attempts to influence friendly populations and military forces situational understanding by providing friendly truth (FM 100-6)

counterterrorism offensive measures taken to prevent, deter, and respond to terrorism (JP 3-07.2)

course of action (Army) a feasible way to accomplish a task or mission that follows the guidance given, will not result in undue damage or risk to the command, and is noticeably different from other actions being considered (FM 101-5)

cover (Army) 1. Protection from the effects of enemy fire. 2. A form of security operation whose primary task is to protect the main body by fighting to gain time while also observing and reporting information and preventing enemy ground observation of and direct fire against the main body. (FM 3-90)

criteria of success information requirements developed during the operations process that measure the degree of success in accomplishing the unit's mission. They are normally expressed as either an explicit evaluation of the present situation or forecast of the degree of mission accomplishment. (FM 6-0)

CSS combat service support

cultural resource monuments, nationally identifiable or distinctive buildings and structures, archives and libraries, ancient artifacts and structures, archaeologically important sites, historically important sites or structures, mosques, cathedrals, temples, other churches or sacred structures, sacred sites or areas, museums, and works of art (FM 41-10)

DA direct action; Department of the Army

D-day the unnamed day on which a particular operation commences or is to commence (JP 5-00.1)

dead space 1. An area within the range of a weapon, radar, or observer, which cannot be covered by fire or observation from a particular position because of intervening obstacles, the nature of the ground, the characteristics of the trajectory, or the limitations of the pointing capabilities of the weapon. 2. An area or zone which is within range of a radio transmitter, but in which a signal is not received. 3. The volume of space above and around a gun or guided missile system into which it cannot fire because of mechanical or electronic limitations. (JP 1-02)

debarkation the unloading of troops, equipment, or supplies from a ship or aircraft (JP 3-35)

deception those measures designed to mislead the enemy by manipulation, distortion, or falsification of evidence to induce the enemy to react in a manner prejudicial to the enemy's interests (JP 3-58)

decision support template a graphic record of wargaming. The decision support template depicts decision points, timelines associated with movement of forces and the flow of the operation, and other key items of information required to execute a specific friendly course of action. (JP 2-01.3)

decisive engagement in land and naval warfare, an engagement in which a unit is considered fully committed and cannot maneuver or extricate itself. In the absence of outside assistance, the action must be fought to a conclusion and either won or lost with the forces at hand. (JP 1-02)

decisive operations those operations that directly accomplish the task assigned by the higher headquarters. They conclusively determine the outcome of major operations, battles, and engagements. (FM 3-0)

decisive point a geographic place, specific key event, or enabling system that allows commanders to gain a marked advantage over an enemy and greatly influences the outcome of an operation (FM 3-0)

defeat a tactical mission task that occurs when an enemy force has temporarily or permanently lost the physical means or the will to fight. The defeated force's commander is unwilling or unable to pursue his adopted course of action, thereby yielding to the friendly commander's will and can no longer interfere to a significant degree with the actions of friendly forces. Defeat can result from the use of force or the threat of its use. (FM 3-90)

defeat in detail defeat in detail is achieved by concentrating overwhelming combat power against separate parts of a force rather than defeating the entire force at once (FM 3-90)

defense in depth the siting of mutually supporting defense positions designed to absorb and progressively weaken attack, prevent initial observations of the whole position by the enemy, and to allow the commander to maneuver the reserve (JP 1-02)

defensive information operations (Army) the integration and coordination of policies and procedures, operations, personnel, and technology to protect and defend friendly information and information systems. Defensive information operations ensure timely, accurate, and relevant information access while denying adversaries the opportunity to exploit friendly information and information systems for their own purposes. (FM 3-0)

defensive operations operations to defeat an enemy attack, buy time, economize forces, or develop conditions favorable for offensive operations. Defensive operations alone normally cannot achieve a decision. Their purpose is to create conditions for a counteroffensive that allows Army forces to regain the initiative. (FM 3-0)

delay a form of retrograde operation in which a force under pressure trades space for time by slowing down the enemy's momentum and inflicting maximum damage on the enemy without, in principle, becoming decisively engaged (JP 1-02)

demonstration (Army) In stability operations and support operations, an operation by military forces in sight of an actual or potential enemy to show military capabilities. (FM 3-07)

depth the extension of operations in time, space, and resources (FM 3-0)

destroy 1. A tactical mission task that physically renders an enemy force combat-ineffective until it is reconstituted. 2. To damage a combat system so badly that it cannot perform any function or be restored to a usable condition without being entirely rebuilt. (FM 3-90)

direct action short-duration strikes and other small-scale offensive actions by special operations forces or special operations-capable units to seize, destroy, capture, recover, or inflict damage on designated personnel or materiel. In the conduct of these operations, special operations forces or special operations-capable units may employ raid, ambush, or direct assault tactics; emplace mines and other munitions; conduct standoff attacks by fire from air, ground, or maritime platforms; provide terminal guidance for precision-guided munitions; conduct independent sabotage; and conduct anti-ship operations. (JP 3-05)

direct approach to apply combat power directly against the enemy center of gravity or the enemy's principal strength (FM 3-0)

direct fire gunfire delivered on a target, using the target itself as a point of aim for either the gun or the director (FM 7-20)

disinformation information disseminated primarily by intelligence organizations or other covert agencies designed to distort information or deceive or influence US decision makers, US forces, coalition allies, key actors or individuals via indirect or unconventional means (FM 100-6)

dislocated civilian (Army) a generic term that describes a civilian who has been forced to move by war, revolution, or natural or man-made disaster from his or her home to some other location. Dislocated citizens include displaced persons, refugees, evacuees, stateless persons, or war victims. Legal and political considerations define the subcategories of a dislocated civilian. (FM 3-07)

displaced person a civilian who is involuntarily outside the boundaries of his or her country or as an internally displaced person is a civilian involuntarily outside their area or region within their country (FM 34-1)

display (Army) representing relevant information in a usable, easily understood audio or visual form tailored to the needs of the user that conveys the common operational picture for decision making and exercising command and control functions (FM 6-0)

disposition distribution of the elements of a command within an area, usually the exact location of each unit headquarters and the deployment of the forces subordinate to it (JP 2-01.3)

disrupt a tactical mission task in which a commander integrates direct and indirect fires, terrain, and obstacles to upset an enemy's formation or tempo, interrupt his timetable, or cause his forces to commit prematurely or attack in a piecemeal fashion. Disrupt is also an engineer obstacle effect that focuses fire planning and obstacle effort to cause the enemy to break up his formation and tempo, interrupt his timetable, commit breaching assets prematurely, and attack in a piecemeal effort. (FM 3-90)

distribution system that complex of facilities, installations, methods, and procedures designed to receive, store, maintain, distribute, and control the flow of military materiel between the point of receipt into the military system and the point of issue to using activities and units (JP 4-0)

DNBI disease and nonbattle injury

doctrine fundamental principles by which the military forces or elements thereof guide their actions in support of national objectives. It is authoritative but requires judgment in application. (JP 1-01)

doctrinal template a model based on known or postulated adversary doctrine. Doctrinal templates illustrate the disposition and activity of adversary forces and assets conducting a particular operation unconstrained by the effects of the battlespace. They represent the application of adversary doctrine under ideal conditions. Ideally, doctrinal templates depict the threat's normal organization for combat, frontages, depths, boundaries and other control measures, assets available from other commands, objective depths, engagement areas, battle positions, and so forth. Doctrinal templates are usually scaled to allow ready use with geospatial products. (JP 2-01.3)

DOD Department of Defense

domestic support operations those activities and measures taken by the Department of Defense to foster mutual assistance and support between the Department of Defense and any civil government agency in planning or preparedness for, or in the application of resources for response to, the consequences of civil emergencies or attacks, including national security emergencies (JP 3-07.7)

EAC echelons above corps

economy one of the eight characteristics of combat service support: providing the most efficient support at the least cost to accomplish the mission (FM 100-10)

economy of force one of the nine principles of war: allocate minimum essential combat power to secondary efforts (FM 3-0)

EEFI essential elements of friendly information

electromagnetic spectrum the range of frequencies of electromagnetic radiation from zero to infinity; it is divided into 26 alphabetically designated bands (JP 3-51)

electronic attack that division of electronic warfare involving the use of electromagnetic energy, directed energy, or antiradiation weapons to attack personnel, facilities, or equipment with the intent of degrading, neutralizing, or destroying enemy combat capability and is considered a form of fires. EA includes: 1. actions taken to prevent or reduce an enemy's effective use of the electromagnetic spectrum, such as jamming and electromagnetic deception, and 2. employment of weapons that use either electromagnetic or directed energy as their primary destructive mechanism (lasers, radio frequency weapons, particle beams), or antiradiation weapons. (JP 3-51)

electronic warfare any military action involving the use of electromagnetic and directed energy to control the electromagnetic spectrum or to attack the enemy. The three major subdivisions within electronic warfare are: electronic attack, electronic protection, and electronic warfare support. (JP 3-51)

electronic warfare support that division of electronic warfare involving actions tasked by, or under direct control of, an operational commander to search for, intercept, identify, and locate or localize sources of intentional and unintentional radiated electromagnetic energy for the purpose of immediate threat recognition, targeting, planning, and conduct of future operations. Thus, electronic warfare support provides information required for immediate decisions involving electronic warfare operations and other tactical actions such as threat avoidance, targeting, and homing. Electronic warfare support data can be used to produce signals intelligence, provide targeting for electronic or destructive attack, and produce measurement and signature intelligence. (JP 3-51)

embarkation the process of putting personnel and/or vehicles and their associated stores and equipment into ships and/or aircraft (JP 3-35)

encirclement an operation where one force loses it freedom of maneuver because an opposing force is able to isolate it by controlling all ground lines of communication (FM 3-0)

end state (Army) a set of required conditions that, when achieved, attain the aims set for the campaign or operation (FM 3-0)

engagement a small, tactical conflict between opposing maneuver forces, usually conducted at brigade level and below (FM 30)

envelopment (Army) a form of maneuver in which an attacking force seeks to avoid the principal enemy defenses by seizing objectives to the enemy rear to destroy the enemy in his current positions. At the tactical level, envelopments focus on seizing terrain, destroying specific enemy forces, and interdicting enemy withdrawal routes. (FM 3-0)

ENY enemy (graphics only)

EOD explosive ordnance disposal

essential elements of friendly information (Army) the critical aspects of a friendly operation that, if known by the enemy, would subsequently compromise, lead to failure, or limit success of the operation, and therefore must be protected from enemy detection (FM 6-0)

event template a guide for collection planning. The event template depicts the named areas of interest where activity, or its lack of activity, will indicate which course of action the adversary has adopted. (JP 2-03.1)

EW electronic warfare

exfiltrate a tactical mission task where a commander removes personnel or units from areas under enemy control by stealth, deception, surprise, or clandestine means (FM 3-90)

exploitation 1. Taking full advantage of success in military operations, and following up initial gains, and making permanent the temporary effects already achieved 2. A type of offensive operation that usually follows a successful attack and is designed to disorganize the enemy in depth. (JP 1-02)

explosive ordnance disposal the detection, identification, on-site evaluation, rendering safe, recovery, and final disposal of unexploded explosive ordnance. It may also include explosive ordnance which has become hazardous by damage or deterioration. (JP 4-04)

FE Far East (graphics only)

firepower (Army) the potential capacity (product) of all weapons and attack systems available to the force commander (FM 3-0)

fires (Army) the delivery of all types of ordnance through both direct and indirect means, as well as nonlethal means, that contribute to the destruction, disruption, or suppression of the enemy; facilitate tactical movement; and achieve a decisive impact (FM 6-20)

fire support battlefield operating system encompasses the collective and coordinated use of target-acquisition data, indirect-fire weapons, fixed-wing aircraft, offensive information operations, and other lethal and nonlethal means against targets located throughout an area of operations (FM 7-15)

fire support coordinating measure a measure employed by land or amphibious commanders to facilitate the rapid engagement of targets and simultaneously provide safeguards for friendly forces (JP 3-09)

fix (Army) a tactical mission task where a commander prevents the enemy from moving any part of his force from a specific location for a specific period of time. Fix is also an engineer obstacle effect that focuses fire planning and obstacle effort to slow an attacker's movement within a specified area, normally an engagement area (FM 3-90)

flexibility one of the eight characteristics of combat service support: being able to adapt combat service support structures and procedures to changing situations, missions, and concepts of operations. (FM 100-10)

FM field manual; frequency modulation

force projection the ability to project the military element of national power from the continental United States (CONUS) or another theater in response to requirements for military operations. Force-projection operations extend from mobilization and deployment of forces, to redeployment to CONUS or home theater. (JP 1)

force protection (Army) those actions taken to prevent or mitigate hostile actions against Department of Defense personnel (to include family members), resources, facilities, and critical information. These actions conserve the force's fighting potential so it can be applied at the decisive time and place and incorporates the coordinated and synchronized offensive and defensive measures to enable the effective employment of the joint force while degrading opportunities for the enemy. Force protection does not include actions to defeat the enemy or protect against accidents, weather, or disease. (FM 3-0)

force tailoring the process of determining the right mix and sequence of units for a mission (FM 3-0)

foreign internal defense participation by civilian and military agencies of a government in any of the action programs taken by another government to free and protect its society from subversion, lawlessness, and insurgency (JP 3-07.1)

forms of maneuver distinct tactical combinations of fire and movement with a unique set of doctrinal characteristics that differ primarily in the relationship between the maneuvering force and the enemy. The choices of offensive maneuver are envelopment, turning movement, infiltration, penetration, and frontal attack. The choices of defensive maneuver are forward and in-depth. Commanders use these to orient on the enemy, not terrain. More than one may be applied during an operation and may be used in conjunction with a form of tactical operation. (FM 3-0)

fratricide the unintentional killing or wounding of friendly personnel by friendly firepower (FM 3-0)

frontal attack (Army) a form of maneuver in which an attacking force seeks to destroy a weaker enemy force or fix a larger enemy force in place over a broad front (FM 3-0)

full spectrum operations the range of operations Army forces conduct in war and military operations other than war (FM 3-0)

GIS geographic information system

GTL gun-target line

guerrilla warfare military and paramilitary operations conducted in enemy-held or hostile territory by irregular, predominantly indigenous forces (JP 3-05)

gun-target line an imaginary straight line from gun to target (JP 3-09)

hazardous material any substance which has a human health hazard associated with it. Special storage, use, handling, and shipment safety procedures and protocols must be followed to help protect against accidental human exposure. Hazardous materials are specifically identified under federal law. (FM 3-100.4)

health threat a composite of ongoing or potential enemy actions; environmental, occupational, and geographic and meteorological conditions; endemic diseases; and employment of nuclear, biological, and chemical weapons (to include weapons of mass destruction) that can reduce the effectiveness of joint forces through wounds, injuries, illness, and psychological stressors (JP 4-02)

high-angle fire (Army) fire delivered to clear an obstacle (such as a hill) that low-angle fire cannot, or fire delivered to attack targets on the reverse side of an obstacle (such as a hill) that cannot be attacked with low-angle or direct fire (FM 6-30)

host nation a nation that receives the forces and/or supplies of allied nations, coalition partners, and/or NATO organizations to be located on, to operate in, or to transit through its territory (JP 3-16)

host-nation support civil and/or military assistance rendered by a nation to foreign forces within its territory during peacetime, crises or emergencies, or war based on agreements mutually concluded between nations (JP 3-16)

HQ headquarters

HRS human resources support

human intelligence a category of intelligence derived from information collected and provided by human sources (JP 2-0)

humanitarian and civic assistance assistance to the local populace provided by predominantly US forces in conjunction with military operations and exercises. This assistance is specifically authorized by Title 10, United States Code, section 401, and funded under separate authorities. Assistance provided under these provisions is limited to (1) medical, dental, and veterinary care provided in rural areas of a country; (2) construction of rudimentary surface transportation systems; (3) well drilling and construction of basic sanitation facilities; and (4) rudimentary construction and repair of public facilities. Assistance must fulfill unit training requirements that incidentally create humanitarian benefit to the local populace. (JP 3-07)

humanitarian assistance programs conducted to relieve or reduce the results of natural or manmade disasters or other endemic conditions such as human pain, disease, hunger, or privation that might present a serious threat to life or that can result in great damage to or loss of property. Humanitarian assistance provided by US forces is limited in scope and duration. The assistance provided is designed to supplement or complement the efforts of the host nation civil authorities or agencies that may have the primary responsibility for providing humanitarian assistance. (JP 3-07.6)

HUMINT human intelligence

HWY highway (graphics only)

IDF Israeli Defense Forces

IED improvised explosive device

imagery intelligence intelligence derived from the exploitation of collection by visual photography, infrared sensors, lasers, electro-optics, and radar sensors, such as synthetic aperture radar, wherein images of objects are reproduced optically or electronically on film, electronic display devices, or other media (JP 2-0)

IMINT imagery intelligence

imitative electromagnetic deception (Army) imitating enemy electromagnetic radiation (predominately communications) through his electromagnetic channels to deceive him or to disrupt his operations (FM 100-6)

indirect fire fire delivered on a target that is not itself used as a point of aim for the weapons or the director (JP 3-09)

infiltration (Army) a form of maneuver in which an attacking force conducts undetected movement through or into an area occupied by enemy forces to occupy a position of advantage in the enemy rear while exposing only small elements to enemy defensive fires (FM 3-0)

information (Army) 1. The meaning assigned to sensing from the environment. 2. On the cognitive hierarchy consists of processed data that provides further meaning with further transformation. Processing activities include filtering, formatting, organizing, collating, correlating, plotting, translating, categorizing, and arranging, among others. (FM 6-0)

information management the provision of relevant information to the right person at the right time in a usable form to facilitate situational understanding and decision making. It uses procedures and information systems to collect, process, store, display, and disseminate information. (FM 3-0)

information operations (Army) the actions taken to affect adversary and influence others' decision-making processes, information, and information systems while protecting one's own information and information systems (FM 3-0)

information requirements (Army) all of the information elements required by the commander and his staff for the successful execution of operations, that is, all elements necessary to address the factors of METT-TC. (FM 6-0)

information superiority (Army) the operational advantage derived from the ability to collect, process, and disseminate an uninterrupted flow of information while exploiting or denying an adversary's ability to do the same (FM 3-0)

information system (Army) the equipment and facilities that collect, process, store, display and disseminate information. This includes computers-hardware and software-and communications, as well as policies and procedures for their use. (FM 3-0)

INFOSYS information systems

initiative (operational) setting or dictating the terms of action throughout the battle or operation (FM 3-0)

insurgency an organized movement aimed at the overthrow of a constituted government through the use of subversion and armed conflict (JP 3-05)

integration one of the eight characteristics of combat service support: the total inclusion of Army combat service support into the operations process (plan, prepare, execute, assess) as well as into other logistic components of the unified force (FM 100-10)

intelligence (Army) the product resulting from the processing and analysis of information collected by any means concerning enemies, potential enemies, or current and potential operational environments (FM 34-1)

intelligence battlefield operating system the activity to generate knowledge of and products portraying the enemy and environmental features required by a commander in planning, preparing, executing, and assessing operations (FM 7-15)

intelligence preparation of the battlefield an analytical methodology employed as part of intelligence planning, to reduce uncertainties concerning the enemy, environment, and terrain for all types of operations. Intelligence preparation of the battlefield is conducted during mission planning to support the commander's decision making and to form the basis for the direction of intelligence operations in support of current and future missions. It utilizes existing databases and identifies gaps in intelligence needed to determine the impact of the enemy, environment, and terrain on operations and presents this in an appropriate form to facilitate operational planning. It forms the basis for situation development. (FM 34-130)

intelligence, surveillance, and reconnaissance the integration and synchronization of all battlefield operating systems to collect and process information about the enemy and environment that produces relevant information to facilitate decision making (FM 3-55)

intelligence, surveillance, and reconnaissance plan an integrated plan for collection of information from all available sources and analysis of that information to produce intelligence to meet requirements. Specifically, a logical plan for transforming priority intelligence requirements (PIR) into orders or requests to reconnaissance and surveillance assets to collect pertinent information within a required time limit. (FM 34-3)

intention an aim or design (as distinct from capability) to execute a specified course of action (JP 1-02)

interagency activities or operations conducted by or through coordination with two or more agencies or an agency and one or more services of the same nation (FM 3-07)

interdict a tactical mission task where the commander prevents, disrupts, or delays the enemy's use of an area or route (FM 3-90)

interior lines a force operates on interior lines when its operations diverge from a central point (FM 3-0)

IO information operations

IPB intelligence preparation of the battlefield

IRA Irish Republican Army

isolate a mission tactical task that requires a unit to seal off—both physically and psychologically—an enemy from his sources of support, deny an enemy freedom of movement, and prevent an enemy unit from having contact with other enemy forces (FM 3-90)

ISR intelligence, surveillance, and reconnaissance

JFC joint force commander

JOA joint operations area

joint force a general term applied to a force composed of significant elements, assigned or attached, of two or more Military Departments, operating under a single joint force commander (JP 3-0)

joint force commander a general term applied to a combatant commander, subunified commander, or joint task force commander authorized to exercise combatant command (command authority) or operational control over a joint force (JP 3-0)

joint operations a general term to describe military actions conducted by joint forces, or by Service forces in relationships (e.g., support, coordinating authority), which, of themselves, do not create joint forces (JP 3-0)

joint task force a joint force that is constituted and so designated by the Secretary of Defense, a combatant commander, a subunified commander, or an existing joint task force commander (JP 3-0)

JP joint publication

JSOTF joint special operations task force

JSTARS Joint Surveillance, Target Attack Radar System

JTF joint task force

JUO joint urban operation

key terrain any locality or area, the seizure or retention of which affords a marked advantage to either combatant in a given course of action (JP 2-01.3)

km kilometers (graphics only)

LAN local area network

law of war that part of international law that regulates the conduct of armed hostilities. Also called the law of armed conflict. (JP 1-04)

liaison that contact or intercommunication maintained between elements of military forces or other agencies to ensure mutual understanding and unity of purpose and action (JP 3-08)

line of communications a route, either land, water, and/or air, which connects an operating military force with a base of operations and along which supplies and military forces move (JP 4-0)

LNO liaison officer

LOC line of communications

LOGCAP logistics civilian augmentation program

logistics the science of planning and carrying out the movement and maintenance of forces. In its most comprehensive sense, those aspects of military operations which deal with: a. design and development, acquisition, storage, movement, distribution, maintenance, evacuation, and disposition of materiel; b. movement, evacuation, and hospitalization of personnel; c. acquisition or construction, maintenance, operation, and disposition of facilities; and d. acquisition or furnishing of services. (JP 4-0)

logistics over-the-shore operations the loading and unloading of ships without the benefit of deep draft-capable, fixed port facilities, in friendly or nondefended territory, and, in time of war, during phases of theater development in which there is no opposition by the enemy; or as a means of moving forces closer to tactical assembly areas dependent on threat force capabilities. (JP 4-01.6)

logistics preparation of the theater all actions taken by combat service support to maximize the means of supporting commander's plans (FM 100-10)

LOS line of sight

LPT logistics preparation of the theater

MACV Military Assistance Command–Vietnam

MAGTF Marine air-ground task force

major operation a series of tactical actions (battles, engagements, strikes) conducted by various combat forces of a single or several services, coordinated in time and place, to accomplish operational, and sometimes strategic objectives in an operational area (FM 3-0)

maneuver (Army) one of the nine principles of war: place the enemy in a disadvantageous position through the flexible application of combat power (FM 3-0)

maneuver battlefield operating system the movement of forces to achieve a position of advantage with respect to enemy forces. This system includes the employment of forces on the battlefield in combination with direct fire or fire potential. This system also includes the conduct of tactical tasks associated with force projection. (FM 7-15)

MANPADS man-portable air defense system

Marine air-ground task force the Marine Corps principal organization for all missions across the range of military operations, composed of forces task-organized under a single commander capable of responding rapidly to a contingency anywhere in the world. The types of forces in the Marine air-ground task force (MAGTF) are functionally grouped into four core elements: a command element, an aviation combat element, a ground combat element, and a combat service support element. The four core elements are categories of forces, not formal commands. The basic structure of the MAGTF never varies, though the number, size, and type of Marine Corps units comprising each of its four elements will always be mission dependent. The flexibility of the organizational structure allows for one or more subordinate MAGTFs to be assigned. (JP 3-02.1)

Marine expeditionary force the largest Marine air-ground task force (MAGTF) and the Marine Corps' principal warfighting organization, particularly for larger crises or contingencies. It is task-organized around a permanent command element and normally contains one or more Marine divisions, Marine aircraft wings, and Marine force service support groups. The Marine expeditionary force is capable of missions across the range of military operations, including amphibious assault and sustained operations ashore in any environment. It can operate from a sea base, a land base, or both. (JP 3-02.1)

Marine expeditionary unit a Marine air-ground task force (MAGTF) that is constructed around an infantry battalion reinforced, a helicopter squadron reinforced, and a task-organized combat service support element. It normally fulfills Marine Corps forward sea-based deployment requirements. The Marine expeditionary unit provides an immediate reaction capability for crisis response and is capable of limited combat operations. (JP 3-02.1)

mass (Army) one of the nine principles of war: concentrate the effects of combat power at the decisive place and time (FM 3-0)

mass casualty any large number of casualties produced in a relatively short period of time, usually as the result of a single incident such as a military aircraft accident, hurricane, flood, earthquake, or armed attack that exceeds local logistical support capabilities (JP 4-02.2)

medical evacuation the timely and efficient movement of patients while providing *en route* medical care to and between medical treatment facilities (FM 4-02)

meeting engagement (Army) a combat action that occurs when a moving force engages an enemy at an unexpected time and place (FM 3-0)

MEF Marine expeditionary force

METT-TC 1) in the context of information management, the major subject categories into which relevant information is grouped for military operations: mission, enemy, terrain and weather, troops and support available, time available, civil considerations (2) in the context of tactics, the major factors considered during mission analysis (FM 6-0)

MEU Marine expeditionary unit

MEU(SOC) Marine expeditionary unit (special operations capable)

MG major general

MHE materials handling equipment

military deception actions executed to deliberately mislead adversary military decision makers as to friendly military capabilities, intentions, and operations, thereby causing the adversary to take specific actions (or inactions) that will contribute to the accomplishment of the friendly mission (JP 3-58)

misinformation unintentionally incorrect information emanating from virtually anyone, for reasons unknown or to solicit a response or interest that is not political or military in origin (FM 100-6)

mission 1. The task, together with the purpose, that clearly indicates the action to be taken and the reason therefore. 2. In common usage, especially when applied to lower military units, a duty assigned to an individual or unit; a task. (JP 1-02)

mission command the conduct of military operations through decentralized execution based upon mission orders for effective mission accomplishment. Successful mission command results from subordinate leaders at all echelons exercising disciplined initiative within the commander's intent to accomplish missions. It requires an environment of trust and mutual understanding. (FM 6-0)

mission orders a technique for completing combat orders to allow subordinates maximum freedom of planning and action to accomplish missions that leave the "how" of mission accomplishment to the subordinate (FM 6-0)

MLRS Multiple Launch Rocket System

mobile defense (Army) a type of defensive operation that concentrates on the destruction or defeat of the enemy through a decisive attack by a striking force (FM 3-0)

mobility operations (Army) those activities that enable a force to move personnel and equipment on the battlefield without delays due to terrain or obstacles (FM 3-34)

mobility corridors areas where a force will be canalized due to terrain constructions. They allow military forces to capitalize on the principles of mass and speed and are therefore relatively free of obstacles. (JP 2-01.3)

mobility/countermobility/survivability battlefield operating system mobility operations preserve the freedom of maneuver of friendly forces. Countermobility operations deny mobility to enemy forces. Survivability operations protect friendly forces from the effects of enemy weapon systems. (FM 7-15)

modified combined obstacle overlay a joint intelligence preparation of the battlespace product used to portray the effects of each battlespace dimension on military operations. It normally depicts militarily significant aspects of the battlespace environment, such as obstacles restricting military movement, key geography, and military objectives. (JP 2-03.1)

movement to contact a type of offensive operation designed to develop the situation and establish or regain contact (FM 3-0)

MSC Military Sealift Command

MST maintenance support team

MTMC Military Traffic Management Command

MTW major theater war

multinational operations a collective term to describe military actions conducted by forces of two or more nations, usually undertaken within the structure of a coalition or alliance (JP 3-16)

N north (graphics only)

NATO North Atlantic Treaty Organization

naval coastal warfare coastal sea control, harbor defense, and port security, executed both in coastal areas outside the United States in support of national policy and in the United States as part of this Nation's defense (JP 3-33)

naval gunfire support fire provided by Navy surface gun systems in support of a unit or units tasked with achieving the commander's objectives. A subset of naval surface fire support. (JP 3-33)

naval surface fire support fire provided by Navy surface gun, missile, and electronic warfare systems in support of a unit or units tasked with achieving the commander's objectives (JP 3-33)

NBC nuclear, biological, and chemical

NCW naval coastal warfare

neutral (Army) an individual, a group of individuals, an organization, or a nation which is not hostile or in any way supportive of only one belligerent force in a hostile environment (FM 3-07)

NGO nongovernmental organization

NIMA National Imagery and Mapping Agency

NKPA North Korean People's Army

no-fire area a land area designated by the appropriate commander into which fires or their effects are prohibited (JP 3-09)

noncombatant 1. An individual, in an area of combat operations, who is not armed and is not participating in any activity in support of any of the factions or forces involved in combat. 2. An individual, such as a chaplain or medical personnel, whose duties do not involve combat. (FM 3-07)

noncombatant evacuation operation an operation directed by the Department of State, the Department of Defense, or other appropriate authority whereby noncombatants are evacuated from foreign countries when their lives are endangered by war, civil unrest, or natural disaster to safe havens or to the United States (JP 3-07.5)

noncontiguous area of operations when one or more of a commander's subordinate forces' areas of operation do not share a common boundary (FM 3-90)

nongovernmental organization a transnational organization of private citizens that maintain a consultative status with the Economic and Social Council of the United Nations. Nongovernmental organizations may be professional associations, foundations, multinational businesses, or simply groups with a common interest in humanitarian assistance activities (development and relief). "Nongovernmental organizations" is a term normally used by non-US organizations. (JP 3-07)

nonlethal fires any fires that do not directly seek the physical destruction of the intended target and are designed to impair, disrupt, or delay the performance of enemy operational forces, functions, and facilities. Psychological operations, special operations forces, electronic warfare (jamming), and other command and control countermeasures are all nonlethal fire options. (FM 6-20)

nonlethal weapons weapons that are explicitly designed and primarily employed so as to incapacitate personnel or material, while minimizing fatalities, permanent injury to personnel, and undesired damage to property and the environment. Unlike conventional lethal weapons that destroy their targets through blast, penetration, and fragmentation, nonlethal weapons employ means other than gross physical destruction to prevent the target from functioning. Nonlethal weapons are intended to have one, or both, of the following characteristics: (1) They have relatively reversible effects on personnel and material. (2) They affect objects differently within their area of influence. (JP 1-02)

NVA North Vietnamese Army

OAKOC observation and fields of fire, avenues of approach, key terrain, obstacles, and cover and concealment

OBJ objective (graphics only)

objective (Army) 1. A location on the ground used to orient operations, phase operations, facilitate changes of direction, and provide for unity of effort. (FM 3-90) 2. One of the nine principles of war:

direct every military operation toward a clearly defined, decisive and attainable objective. (FM 3-0) 3. The most important decisive point. (FM 3-0)

offensive one of the nine principles of war: seize, retain, and exploit the initiative (FM 3-0)

offensive information operations (Army) the integrated use of assigned and supporting capabilities and activities, mutually supported by intelligence, to affect enemy decision makers or to influence others to achieve or promote specific objectives (FM 3-0)

offensive operations operations aimed at destroying or defeating an enemy. Their purpose is to impose US will on the enemy and achieve decisive victory. (FM 3-0)

operation 1. A military action or the carrying out of a strategic, operational tactical, service, training, or administrative military mission. 2. The process of carrying on combat, including movement, supply, attack, defense, and maneuvers needed to gain the objectives of any battle or campaign. (JP 5-0)

operational control transferable command authority that may be exercised by commanders at any echelon at or below the level of combatant command. Operational control is inherent in combatant command (command authority). Operational control may be delegated and is the authority to perform those functions of command over subordinate forces involving organizing and employing commands and forces, assigning tasks, designating objectives, and giving authoritative direction necessary to accomplish the mission. Operational control includes authoritative direction over all aspects of military operations and joint training necessary to accomplish missions assigned to the command. Operational control should be exercised through the commanders of subordinate organizations. Normally this authority is exercised through subordinate joint force commanders and Service and/or functional component commanders. Operational control normally provides full authority to organize commands and forces and to employ those forces as the commander in operational control considers necessary to accomplish assigned missions. Operational control does not, in and of itself, include authoritative direction for logistics or matters of administration, discipline, internal organization, or unit training. (JP 3-0)

operational framework the arrangement of friendly forces and resources in time, space, and purpose with respect to each other and the enemy or situation. It consists of the area of operations, battlespace, and the battlefield organization. (FM 3-0)

operational level of war the level of war at which campaigns and major operations are planned, conducted, and sustained to accomplish strategic objectives within theaters or operational areas. Activities at this level link tactics and strategy by establishing operational objectives needed to accomplish the strategic objectives, sequencing events to achieve the operational objectives, initiating actions, and applying resources to bring about and sustain these events. These activities imply a broader dimension of time or space than do tactics; they ensure the logistic and administrative support of tactical forces, and provide the means by which tactical successes are exploited to achieve strategic objectives. (JP 3-0)

operations process plan, prepare, and execute with continuous assessment. (FM 6-0)

operations security a process of identifying critical information and subsequently analyzing friendly actions attendant to military operations and other activities to: a. identify those actions that can be observed by adversary intelligence systems; b. Determine indicators hostile intelligence systems might obtain that could be interpreted or pieced together to derive critical information in time to be useful to adversaries; and c. Select and execute measures that eliminate or reduce to an acceptable level the vulnerabilities of friendly actions to adversary exploitation. (JP 3-54)

OPSEC operations security

order (Army) a written or oral communication directing actions. Orders are based on plans or the receipt of a new mission. (FM 101-5)

PA public affairs

paramilitary force forces or groups distinct from the regular armed forces of any country, but resembling them in organization, equipment, training, or mission (JP 3-05)

patrol a detachment of ground, sea, or air forces sent out for the purpose of gathering information or carrying out a destructive, harassing, mopping-up, or security mission (JP 1-02)

PDF Panamanian Defense Force

peace enforcement application of military force, or the threat of its use, normally pursuant to international authorization, to compel compliance with resolutions or sanctions designed to maintain or restore peace and order (JP 3-07.3)

peacekeeping military operations undertaken with the consent of all major parties to a dispute, designed to monitor and facilitate implementation of an agreement (ceasefire, truce, or other such agreement) and support diplomatic efforts to reach a long-term political settlement (JP 3-07.3)

peace operations a broad term that encompasses peacekeeping operations and peace enforcement operations conducted in support of diplomatic efforts to establish and maintain peace (JP 3-07.3)

peacetime military engagement all military activities that involve other nations and are intended to shape the security environment in peacetime. It includes programs and exercises that the US military conducts with other nations to shape the international environment, improve mutual understanding with other countries, and improve interoperability with treaty partners or potential coalition partners. Peacetime military engagement activities are designed to support a combatant commander's objectives as articulated in the theater engagement plan. (FM 3-0)

penetration a form of maneuver in which an attacking force seeks to rupture enemy defenses on a narrow front to disrupt the defensive system (FM 3-0)

PGM precision-guided munitions

phase line a line utilized for control and coordination of military operations, usually an easily identified feature in the operational area (JP 1-02)

physical destruction the application of combat power to destroy or degrade adversary forces, sources of information and command and control systems, and installations. It includes direct and indirect forces from ground, sea, and air forces. Also included are direct actions by special operations forces. (FM 100-6)

physical security that part of security concerned with physical measures designed to safeguard personnel; to prevent unauthorized access to equipment, installations, material, and documents; and to safeguard them against espionage, sabotage, damage, and theft (JP 3-13)

PIR priority information requirements

PL phase line (graphics only)

PLO Palestine Liberation Organization

PMCS preventive maintenance checks and services

PME peacetime military engagement

port of debarkation the geographic point at which cargo or personnel are discharged. May be a seaport or aerial port of debarkation; for unit requirements, it may or may not coincide with the destination (JP 3-35)

port of embarkation the geographic point in a routing scheme from which cargo or personnel depart. This may be a seaport or aerial port from which personnel and equipment flow to port of debarkation; for unit and nonunit requirements, it may or may not coincide with the origin. (JP 3-35)

POW prisoner of war

principles of war principles that provide general guidance for conducting war and military operations other than war at the strategic, operational, and tactical levels. The principles are the enduring bedrock of US military doctrine. The nine principles of war are: objective, offensive, mass, economy of force, maneuver, unity of command, security, surprise, and simplicity. (FM 3-0)

priority information requirements those intelligence requirements for which a commander has an anticipated and stated priority in his task of planning and decision-making (JP 2-0)

propaganda	any form of communication in support of national objectives designed to influence the opinions, emotions, attitudes, or behavior of any group in order to benefit the sponsor, either directly or indirectly (JP 3-53)
psychological operations	planned operations to convey selected information and indicators to foreign audiences to influence their emotions, motives, objective reasoning, and ultimately the behavior of foreign governments, organizations, groups, and individuals. The purpose of psychological operations is to induce or reinforce foreign attitudes and behavior favorable to the originator's objectives. (JP 3-53)
PSYOP	psychological operations
public affairs	those public information, command information, and community relations activities directed toward both the external and internal with interest in the Department of Defense (JP 3-61)
pursuit	an offensive operation designed to catch or cut off a hostile force attempting to escape, with the aim of destroying it (JP 1-02)
push	in logistics, the delivery of a predetermined amount of supplies to a user on a scheduled basis without the user requesting them. (FM 100-10)
PZ	Panzer (graphics only)
QRF	quick reaction force
R.	River (graphics only)
railhead	a point on a railway where loads are transferred between trains and other means of transport (JP 4-01.5)
rear area	for any particular command, the area extending forward from its rear boundary to the rear of the area assigned to the next lower level of command. This area is provided primarily for the performance of support functions and is where the majority of the echelon's sustaining functions occur. (JP 3-10)
refugee	a civilian who, by reason of real or imagined danger, has left home to seek safety elsewhere (JP 3-07.6)
relevant information	all information of importance to commanders and staffs in the exercise of command and control (FM 3-0)
relief in place	(Army) a tactical enabling operation in which, by direction of higher authority, all or part of a unit is replaced in an area by the incoming unit (FM 3-90)
reorganization	action taken to shift internal resources within a degraded unit to increase its level of combat effectiveness (FM 100-9)
responsiveness	one of the eight characteristics of combat service support: providing the right support at the right place at the right time (FM 100-10)
reserve	that portion of a force withheld from action or uncommitted to a specific course of action, so as to be available for commitment at

the decisive moment. Its primary purpose is to retain flexibility through offensive action. (JP 1-02)

retrograde a type of defensive operations that involves organized movement away from the enemy (FM 3-0)

riot control agent any chemical that is not listed in the Chemical Weapons Convention, which can produce rapidly in humans sensory irritate or disabling physical effects which disappear within a short time following termination of exposure (JP 1-02)

risk (Army) chance of hazard or bad consequences; exposure to chance of injury or loss. Risk level is expressed in terms of hazard probability or severity. (FM 100-14)

risk assessment the identification and assessment of hazards (first two steps of risk management process) (JP 1-02)

risk management (Army) the process of identifying, assessing, and controlling risk arising from operational factors, and making an informed decision that balances cost with mission benefits (FM 3-0)

ROE rules of engagement

ROK Republic of Korea

ROM Romanian Army (graphics only)

RPG rocket-propelled grenade

rules of engagement (Joint) directives issued by competent military authority which delineate the circumstances and limitations under which United States forces will initiate and/or continue combat engagement with other forces encountered (JP 3-0)

SALT supporting arms liaison team

SBU special boat unit

SEAL sea-air-land team

search and attack a technique for conducting a movement to contact that shares many of the characteristics of an area security mission (FM 3-0)

secure (Army) a tactical mission task that involves preventing a unit, facility, or geographical location from being damaged or destroyed as a result of enemy action. (FM 3-90)

security (Army) one of the nine principles of war: never permit the enemy to acquire an unexpected advantage (FM 3-0)

seize a tactical mission task that involves taking possession of a designated area using overwhelming force (FM 3-90)

sequel an operation that follows the current operation. It is a future operation that anticipates the possible outcome—success, failure, or stalemate—of the current operation. (FM 3-0)

SF Army Special Forces

shaping operations operations at any echelon that create and preserve conditions for success of the decisive operation (FM 3-0)

show of force an operation, designed to demonstrate US resolve that involves increased visibility of US deployed forces in an attempt to defuse a specific situation, that, if allowed to continue, may be detrimental to US interests or national objectives (JP 3-07)

SIGINT signals intelligence

signals intelligence 1. A category of intelligence comprising either individually or in combination all communications intelligence, electronic intelligence, and foreign instrumentation signals intelligence, however transmitted. 2. Intelligence derived from communications, electronic, and foreign instrumentation signals. (JP 2-0)

simplicity one of the nine principles of war and one of the eight characteristics of combat service support: prepare clear, uncomplicated plans, and clear, concise orders to ensure thorough understanding. (FM 3-0)

situational understanding the product of applying analysis and judgment to the common operational picture to determine the relationships among the factors of METT-TC (FM 3-0)

situation template a depiction of assumed adversary dispositions, based on adversary doctrine and the effects of the battlespace if the adversary should adopt a particular course of action. In effect, situation templates are the doctrinal templates depicting a particular operation modified to account for the effects of the battlespace environment and the adversary's current situation (training and experience levels, logistic status, losses, dispositions). Normally, the situation template depicts adversary units two levels of command below the friendly force, as well as the expected locations of high-value targets. Situation templates use time-phase lines to indicate movement of forces and the expected flow of the operation. Usually, the situation template depicts a critical point in the course of action. Situation templates are one part of an adversary course of action model. Models may contain more than one situation template. (JP 2-03.1)

SJA Staff Judge Advocate

SO special operations

SOCCE special operations command and control element

SOCOORD special operations coordination element

SOF special operations forces

SOFA status-of-forces agreement

SOP standing operating procedure

SOWT special operations weather team

SPOD seaport of debarkation

SPOE seaport of embarkation

space operations the employment of space system capabilities that provide the means to enhance command and control, facilitate the maneuver of forces, reduce the commander's uncertainty, and improve fire support, air defense, intelligence collection, and combat service support operations which will support strategic, operational, and tactical missions across the operational continuum in the near, mid, and far term (FM 3-14)

special operations (Joint) operations conducted by specially organized, trained, and equipped military and paramilitary forces to achieve military, political, economic, or informational objectives by unconventional military means in hostile, denied, or politically sensitive areas. These operations are conducted across the full range of military operations, independently or in coordination with operations of conventional, nonspecial operations forces. Political-military considerations frequently shape special operations, requiring clandestine, covert, or low visibility techniques and oversight at the national level. Special operations differ from conventional operations in degree of physical and political risk, operational techniques, mode of employment, independence from friendly support, and dependence on detailed operational intelligence and indigenous assets. (JP 3-05)

special operations forces those Active and Reserve Component forces of the Military Services designated by the Secretary of Defense and specifically organized, trained, and equipped to conduct and support special operations (JP 3-05)

special reconnaissance (Army) complementing of national and theater intelligence collection assets and systems by obtaining specific, well-defined, and time-sensitive information of strategic or operational significance. It may complement other collection methods where there are constraints of weather, terrain-masking, hostile countermeasures, and/or other systems availability. Special reconnaissance is a human intelligence function that places US or US-controlled "eyes on target" in hostile, denied, or politically sensitive territory when authorized. SOF may conduct these missions unilaterally or in support of conventional operations (FM 100-25)

SR special reconnaissance

SSC smaller-scale contingency

stability operations operations that promote and protect US national interests by influencing the threat, political, and information dimensions of the operational environment through a combination of peacetime developmental, cooperative activities and coercive actions in response to crisis (FM 3-0)

status-of-forces agreement an agreement that defines the legal position of a visiting military force deployed in the territory of a friendly state. Agreements delineating the status of visiting military forces may be bilateral or multilateral. Provisions pertaining to the status of visiting forces may be set forth in a separate agreement, or they may form a part of a more comprehensive agreement. These provisions describe how the authorities of a visiting force may control members of that force and the amenability of the force or its members to the local law or to the authority of local officials. To the extent that agreements delineate matters affecting the relations between a military force and civilian authorities and population, they may be considered as civil affairs agreements. (JP 3-57)

strategic level of war the level of war at which a nation, often as a member of a group of nations, determines national or multinational (alliance or coalition) strategic security objectives and guidance, and develops and uses national resources to accomplish these objectives. Activities at this level establish national and multinational military objectives; sequence initiatives; define limits and assess risks for the use of military and other instruments of national power; develop global plans or theater war plans to achieve these objectives; and provide military forces and other capabilities in accordance with strategic plans. (JP 3-0)

strategy the art and science of developing and employing armed forces and other instruments of national power in a synchronized fashion to secure national or multinational objectives (FM 3-0)

striking force a committed force organized to conduct the decisive attack in a mobile defense. It normally comprises the maximum combat power available to the commander at the time of the attack. (FM 3-0)

strong point (Army) a heavily fortified battle position tied to a natural or reinforcing obstacle to create an anchor for the defense or to deny the enemy decisive or key terrain (FM 3-90)

STT special tactics team

support operations operations that employ Army forces to assist civil authorities, foreign or domestic, as they prepare for or respond to crisis and relieve suffering (FM 3-0)

surprise one of the nine principles of war: strike the enemy at a time or place or in a manner for which he is unprepared (FM 3-0)

survivability (Joint) 1. Concept which includes all aspects of protecting personnel, weapons, and supplies while simultaneously deceiving the enemy. Survivability tactics include building a good defense; employing frequent movement; using concealment, deception, and camouflage; and constructing fighting and protective positions for both individuals and equipment. Encompasses planning and locating position sites, designing adequate overhead cover, analyzing terrain conditions and construction materials, selecting excavation methods, and countering the effects of direct and indirect fire weapons. (JP 3-34) (Army) 2. One of the eight characteristics of combat service support: being able to shield support functions from destruction of degradation (FM 100-10)

sustainability one of the eight characteristics of combat service support: the ability to maintain continuous support throughout all phases of the operations (FM 100-10)

sustaining operations operations at any echelon that enable shaping and decisive operations by providing combat service support, rear area and base security, movement control, terrain management, and infrastructure development (FM 3-0)

sustainment the provision of personnel, logistic, and other support required to maintain and prolong operations or combat until successful accomplishment or revision of the mission or of the national objective (JP 4-0)

synchronization 1. The arrangement of military actions in time, space, and purpose to produce maximum relative combat power at a decisive place and time. 2. In the intelligence context, application of intelligence sources and methods in concert with the operational plan. (JP 3-0)

TAC tactical (graphics only)

TACON tactical control

tactical combat force a combat unit, with appropriate combat support and combat service support assets, that is assigned the mission of defeating Level II threats (JP 3-10.1)

tactical control command authority over assigned or attached forces or commands, or military capability or forces made available for tasking, that is limited to the detailed and, usually, local direction and control of movements or maneuvers necessary to accomplish missions or tasks assigned. Tactical control is inherent in operational control. Tactical control may be delegated to, and exercised at any level at or below the level of combatant command. (JP 3-0)

tactical level of war the level of war at which battles and engagements are planned and executed to accomplish military objectives assigned to tactical units or task forces. Activities at this level focus on the ordered arrangement and maneuver of combat elements in relation to each other and to the enemy to achieve combat objectives. (JP 3-0)

target acquisition the detection, identification, and location of a target in sufficient detail to permit the effective employment of weapons (JP 3-55)

targeting 1. The process of selecting targets and matching the appropriate response to them, taking account of operational requirements and capabilities. 2. The analysis of enemy situations relative to the commander's mission, objectives, and capabilities at the commander's disposal, to identify and nominate specific vulnerabilities that, if exploited, will accomplish the commander's purpose through delaying, disrupting, disabling, or destroying enemy forces or resources critical to the enemy. (JP 3-60)

task organization (Army) a temporary grouping of forces designed to accomplish a particular mission (FM 3-0)

task organizing (Army) the process of allocating available assets to subordinate commanders and establishing their command and support relationships (FM 3-0)

TC training circular

TCC transportation component command

tempo the rate of military action (FM 3-0)

terrorism the calculated use of unlawful violence or threat of unlawful violence to inculcate fear; intended to coerce or to intimidate governments or societies in the pursuit of goals that are generally political, religious, or ideological. (JP 3-07.2)

terrorist an individual who uses violence, terror, and intimidation to achieve a result (JP 3-07.2)

terminal control 1. The authority to direct the maneuver of aircraft which are delivering ordnance, passengers, or cargo to a specific location or target. Terminal control is a type of air control. 2. Any electronic, mechanical, or visual control given to aircraft to facilitate target acquisition and resolution. (JP 1-02)

TF task force

throughput distribution the bypassing of one or more intermediate supply echelons in the supply system to avoid multiple handling (FM 100-10)

TIM toxic industrial materials

time-phased force and deployment data a Joint Operation Planning and Execution System database located at Appendix 1 to Annex A of deliberate plans. It identifies types and/or actual units required to support the operation plan and indicates origin and ports of debarkation or ocean area. This listing is to include both a. In-place units; and b. Units to be deployed to support the deliberate plan. (JP 4-0)

track 1. To display or record the successive positions of a moving object. 2. To keep a gun properly aimed, or to point continuously a target-locating instrument at a moving target (JP 1-02)

TRADOC Training and Doctrine Command

TTP tactics, techniques, and procedures

turning movement (Army) a form of maneuver in which the attacking force seeks to avoid the enemy's principal defensive positions by seizing objectives to the enemy rear and causing the enemy to move out of his current positions or divert major forces to meet the threat (FM 3-0)

UAV unmanned aerial vehicle

UN United Nations

unconventional warfare a broad spectrum of military and paramilitary operations, normally of long duration, predominantly conducted by indigenous or surrogate forces who are organized, trained, equipped, supported, and directed in varying degrees by an external source. It includes guerrilla warfare and other direct offensive, low visibility, covert, or clandestine operations, as well as the indirect activities of subversion, sabotage, intelligence activities, and evasion and escape. (JP 3-05)

unexploded ordnance explosive ordnance which has been primed, fused, armed, or otherwise prepared for action, and which has been fired, dropped, launched, projected, or placed in such a manner as to constitute a hazard to operations, installations, personnel, or material, and remains unexploded either by malfunction or design or for any other cause (JP 1-02)

UNITAF Unified Task Force

unity of command one of the nine principles of war: for every objective, ensure unity of effort under one responsible commander (FM 3-0)

unity of effort coordination and cooperation among all forces toward a commonly recognized objective, even if the forces are not necessarily part of the same command structure (FM 6-0)

UNOSOM UN Operations in Somalia

UNSCR UN Security Council Resolution

***UO** urban operations

***urban area** a topographical complex where man-made construction or high population density is the dominant feature

***urban environment** includes the physical urban area as well as the complex and dynamic interaction and relationships between its key components—the terrain (natural and man-made), the population, and the supporting infrastructure—as an overlapping and interdependent system of systems.

urban operations (Army) offense, defense, stability, and support operations conducted in a topographical complex and adjacent natural terrain where manmade construction and high population density are the dominant features (FM 3-0)

US United States

USAF United States Air Force

USAID US Agency for International Development

USCG US Coast Guard

USCINCCENT Commander in Chief, United States Central Command

USFORSOM US Forces, Somalia

USN United States Navy

USS US ship

USSOCOM US Special Operations Command

USSPACECOM US Space Command

USTRANSCOM US Transportation Command

UXO unexploded ordnance

VC Vietcong

versatility the ability of Army forces to meet the global, diverse mission requirements of full spectrum operations (FM 3-0)

visualizing creating and thinking in mental images. See also *commander's visualization*. (FM 6-0)

vulnerability 1. The susceptibility of a nation or military force to any action by any means through which its war potential or combat effectiveness may be reduced or its will to fight diminished. 2. The characteristics of a system that cause it to suffer a definite degradation (incapability to perform the designated mission) as a result of having been subjected to a certain level of effects in an unnatural (manmade) hostile environment. 3. In information operations, a weakness in information system security design, procedures, implementation, or internal controls that could be exploited to gain unauthorized access to information or an information system. (JP 1-02)

weapons of mass destruction weapons that are capable of a high order of destruction and/or of being used in such a manner as to destroy large numbers of people. Weapons of mass destruction can be high explosives or nuclear, biological, chemical, and radiological weapons, but excludes the means of transporting or propelling the weapon where such means is a separable and divisible part of the weapon. (JP 3-11)

withdrawal operation a planned retrograde operation in which a force in contact disengages from an enemy force and moves in a direction away from the enemy (Joint)

WMD weapons of mass destruction

Bibliography

When a field manual has been published under a new number, the bibliography shows the old number after the new number.

DOCUMENTS NEEDED

These documents must be available to the intended users of this publication.

JP 1-02. *Department of Defense Dictionary of Military and Associated Terms.* Available at http://www.dtic.mil/doctrine/jel/new_pubs/jp1_02.pdf

JP 3-06. *Doctrine for Joint Urban Operations.* 16 September 2002.

FM 3-0 (100-5). *Operations.* 14 June 2001.

FM 3-06.11 (90-10-1). *Combined Arms Operations in Urban Terrain.* 28 February 2002.

FM 3-07 (100-20). *Stability Operations and Support Operations.* 20 February 2003.

FM 3-90. *Tactics.* 04 July 2001.

FM 34-130. *Intelligence Preparation of the Battlefield.* 08 July 1994. FM 34-130 will be republished as FM 2-01.3.

FM 100-14. *Risk Management.* 23 April 1998. FM 100-14 will be republished as FM 5-19.

FM 101-5. *Staff Organizations and Operations.* 31 May 1997. FM 101-5 will be republished as FM 5-0 and retitled *Army Planning and Orders Production.*

FM 101-5-1. *Operational Terms and Graphics.* 30 September 1997. FM 101-5-1 will be republished as FM 1-02.

SOURCES USED

These are the sources quoted or paraphrased in this publication.

DEPARTMENT OF DEFENSE PUBLICATIONS

Quadrennial Defense Review Report. 30 September 2001.

JOINT PUBLICATIONS

Most joint publications are available: http://www.dtic.mil/doctrine/jel/

JP 3-0. *Doctrine for Joint Operations.* 10 September 2001.

JP 3-05. *Doctrine for Joint Special Operations.* 17 April 1998.

JP 3-08. *Interagency Coordination During Joint Operations* (2 Volumes). 9 October 1996.

JP 3-33. *Joint Force Capabilities*. 13 October 1999.

JP 4-0. *Doctrine for Logistic Support of Joint Operations*. 06 April 2000.

JP 5-00.2. *Joint Task Force Planning Guidance and Procedures*. 13 January 1999.

Joint Military Operations Historical Collection. 15 July 1997.

ARMY PUBLICATIONS

Most Army doctrinal publications are online: http://155.217.58.58/atdls.htm

FM 1 (100-1). *The Army*. 14 June 2001.

FM 3-05.20 (31-20). *Special Forces Operations*. 26 June 2001.

FM 3-05.30 (33-1). *Psychological Operations*. 19 June 2000.

FM 3-06.1 (1-130). *Aviation Urban Operations: Multiservice Procedures for Aviation Urban Operations*. 15 April 2001.

FM(M) 3-3. *Chemical and Biological Contamination Avoidance*. 16 November 1992. FM(M) 3-3 will be republished as FM(M) 3-11.3.

FM 3-14. *Nuclear, Biological, and Chemical (NBC) Vulnerability Analysis*. 12 November 1997. FM 3-14 will be republished as FM 3-11.14.

FM 3-19.4 (19-4). *Military Police Leaders' Handbook*. 04 March 2002.

FM 3-19.40 (19-40). *Military Police Internment/Resettlement Operations*. 01 August 2001.

FM 3-21. *Chemical Accident Contamination Control*. 23 February 1978. FM 3-21 will be republished as FM 3-11.21.

FM 3-31. *Joint Force Land Component Commander Handbook (JFLCC)*. 13 December 2001.

FM 3-31.1 (90-31). *AMCI: Army and Marine Corps Integration in Joint Operations*. 21 November 2001.

FM 3-34.230 (5-105). *Topographic Operations*. 03 August 2000.

FM 3-55. *Intelligence, Surveillance, and Reconnaissance*. TBP.

FM 3-61.1. *Public Affairs Tactics, Techniques, and Procedures*. 01 October 2000.

FM 3-97.6 (90-6). *Mountain Operations*. 28 November 2000.

FM 5-33. *Terrain Analysis*. 11 July 1990. FM 5-33 will be republished as FM 3-34.330.

FM 5-415. *Fire-Fighting Operations*. 09 February 1999. FM 5-415 will be republished as FM 3-34.480.

FM 6-0. *Command and Control*. TBP.

FM 6-22.5 (22-9). *Combat Stress*. 23 June 2000.

FM 7-15. *The Army Universal Task List*. TBP.

FM 8-51. *Combat Stress Control in a Theater of Operations, Tactics, Techniques, and Procedures*. 30 January 1998. FM 8-51 will be republished as FM 4-02.51.

FM 19-15. *Civil Disturbances*. 25 November 1985. FM 19-15 will be republished as FM 3-19.15.

FM 22-51. *Leaders' Manual for Combat Stress Control*. 29 September 1994. FM 22-51 will be republished as FM 4-02.22.

FM 22-100. *Army Leadership: Be, Know, Do*. 31 August 1999. FM 22-100 will be republished as FM 6-22.

FM 27-10. *The Law of Land Warfare*. 18 July 1956. FM 27-10 will be republished as FM 1-04.10.

FM 27-100. *Legal Support to Operations*. 01 March 2000. FM 27-100 will be republished as FM 1-04.

FM 31-20-3. *Foreign Internal Defense Tactics, Techniques, and Procedures for Special Forces*. 20 September 1994. FM 31-20-3 will be republished as FM 3-05.202.

FM 31-50. *Combat in Fortified and Built-Up Areas*. March 1964.

FM 33-1-1. *Psychological Operations Techniques and Procedures*. 05 May 1994. FM 33-1-1 will be republished as FM 3-05.301.

FM 34-3. *Intelligence Analysis*. 15 March 1990. FM 34-3 will be republished as FM 2-33.4.

FM 34-7. *Intelligence and Electronic Warfare Support to Low Intensity Conflict Operations*. 18 May 1993. FM 34-7 will be republished as FM 2-91.1 and retitled *Intelligence and Electronic Warfare for Stability Operations and Support Operations*.

FM 34-81. *Weather Support for Army Tactical Operations*. 31 August 1989. FM 34-81 will be republished as FM 2-33.2.

FM 34-81-1. *Battlefield Weather Effects*. 23 December 1992. FM 34-81-1 will be republished as FM 2-33.201.

FM 41-10. *Civil Affairs Operations*. 14 February 2000. FM 41-10 will be republished as FM 3-05.40.

FM 44-8. *Combined Arms for Air Defense*. 01 June 1999. FM 44-8 will be republished as FM 3-01.8.

FM 46-1. *Public Affairs Operations*. 30 May 1997. FM 46-1 will be republished as FM 3-61.

FM 90-8. *Counterguerrilla Operations*. 29 August 1986. FM 90-8 will be republished as FM 3-07.11.

FM 90-10. *Military Operations on Urbanized Terrain (MOUT) (How to Fight)*. 15 August 1979.

FM 90-40. *NLW: Multiservice Procedures for the Tactical Employment of Nonlethal Weapons*. 06 October 1998. FM 90-40 will be republished as FM 3-22.40.

FM 90-41. *JTF Liaison Handbook: Multiservice Tactics, Techniques, and Procedures for Joint Task Force (JTF) Liaison Operations*. 26 August 1998. FM 90-41 will be republished as FM 6-14.

FM 100-5. *Operations*. 1944.

FM 100-6. *Information Operations*. 27 August 1996. FM 100-6 will be republished as FM 3-13.

FM 100-8. *The Army in Multinational Operations*. 24 November 1997. FM 100-8 will be republished as FM 3-16.

FM 100-10. *Combat Service Support*. 03 October 1995. FM 100-10 will be republished as FM 4-0.

FM 100-25. *Doctrine for Army Special Operations Forces*. 01 August 1999. FM 100-25 will be republished as FM 3-05 and retitled *Army Special Operations Forces Doctrine*.

TC 90-1. *Training for Urban Operations*. 01 April 2002.

Center for Army Lessons Learned (CALL) U.S. Army Combined Arms Command (CAC). *Operation Restore Hope Lessons Learned Report*. Fort Leavenworth, KS: U.S. Government Printing Office, 1993.

NONMILITARY PUBLICATIONS

26th Infantry Regimental Association. *Aachen: Military Operations in Urban Terrain*. Lititz, PA: 26th Infantry Regimental Association, 1999.

Adan, Abraham. *On the Banks of the Suez*. Novato, CA: Presidio Press, 1980.

Akers, Frank H., Jr. and George B. Singleton. "Task Force Ranger: A Case Study Examining the Application of Advanced Technologies in Modern Urban Warfare." Oak Ridge, TN: National Security Program Office, November 2000.

Allard, Kenneth. *Somalia Operations: Lessons Learned*. Washington, DC: National Defense University Press, 1995.

Bass, Bernard M. "Leading in the Army After Next." *Military Review* (March/April 1998): 46–58.

Beevor, Anthony. *Stalingrad*. New York: Penguin Books, 1998.

Bowden, Mark. *Black Hawk Down: A Story of Modern War*. New York: Atlantic Monthly Press, 1999.

Butts, Kent H. "The Strategic Importance of Water." *Parameters* (spring 1997): 65–83.

Celeski, Joseph D. "Joint Urban Operations: Special Forces in Urban Campaign Planning." *Special Warfare* (fall 2001): 2–13.

Chuikov, Vasili I. *The Battle for Stalingrad*. New York: Holt, Rinehart, and Winston, 1964.

Colby, John. *War From the Ground Up: The 90th Division in WWII.* Austin, TX: Nortex Press, 1991.

Cole, Ronald H. *Operation Just Cause.* Washington, DC: Joint History Office, 1995.

Connaughton, Richard, John Pimlott, and Duncan Anderson. *The Battle for Manila: The Most Devastating Untold Story of World War II.* Novata, CA: Presidio, 1995.

Delk, James D. *Fires & Furies: The LA Riots.* Palm Springs, CA: Etc. Publications, 1995.

Dewar, Michael. *British Army in Northern Ireland.* New York: Sterling Publishers, 1985.

Donnelly, Thomas, Margaret Roth, and Caleb Baker. *Operation Just Cause: The Storming of Panama.* New York: Lexington Books, 1991.

Dupuy, R. Ernest and Trevor N. Dupuy. *The Encyclopedia of Military History: From 3500 B.C. to the Present.* San Francisco: Harper & Row, 1977.

Dupuy, Trevor N. and Paul Martell. *Flawed Victory: The Arab-Israeli Conflict and the 1982 War in Lebanon.* Fairfax, VA: Hero Books, 1986.

Edwards, Sean J. A. *Freeing Mercury's Wings: Improving Tactical Communications in Cities.* Santa Monica, CA: RAND, 2001.

Fishel, John T. *The Fog of Peace: Planning and Executing the Restoration of Panama.* Carlisle, PA: Strategic Studies Institute, 1992.

Funston, Frederick. "How the Army Worked to Save San Francisco." *Cosmopolitan Magazine* (July 1906). Available at http://www.sfmuseum.org/1906/cosmo.html

Gabriel, Richard A. *Operation Peace for Galilee: The Israeli-PLO War in Lebanon.* New York: Hill and Wang, 1984.

Glenn, Russell W. *An Attack on Duffer's Downtown.* Santa Monica, CA: RAND, 2001.

———. *Heavy Matter: Urban Operations' Density of Challenges.* Santa Monica, CA: RAND, 2000.

———. "Urban Combat is Complex." *Proceedings* (February 2002): 62–65.

———. *"...We Band of Brothers": The Call for Joint Urban Operations Doctrine.* Santa Monica, CA: RAND, 1999.

Grau, Lester W. and Timothy Smith. *A 'Crushing' Victory: Fuel-Air Explosives and Grozny 2000.* Fort Leavenworth, KS: Foreign Military Studies Office, 2000.

Hamill, Desmond. *Pig in the Middle: The Army in Northern Ireland, 1969–1984.* London: Methuen, 1985.

Hammel, Eric. *Fire in the Streets: The Battle for Hue, Tet 1968.* New York: Dell Publishing, 1991.

Heinl, Robert Debs, Jr. *Victory at High Tide: The Inchon-Seoul Campaign.* Annapolis, MD: Nautical & Aviation Publishing Company of America, 1979.

Herzog, Cham. *The War of Atonement, October, 1973.* Boston: Little, Brown and Company, 1975.

Higham, Robin. *Bayonets in the Streets: The Use of Troops in Civil Disturbances.* Lawrence, KS: University Press of Kansas, 1969.

Hoar, Joseph P. "A CINC's Perspective." *JFQ Forum* (autumn 1993): 56–63.

Infantry in Battle. 2nd ed. Garret & Massie: Richmond, VA, 1939. Reprint, Fort Leavenworth, KS: US Army Command and General Staff College, 1981.

Jomini, Antoine-Henri. "The Art of War" in *Roots of Strategy: Book 2.* Harrisburg, PA: Stackpole Books, 1987.

Kemp, Anthony. *The Unknown Battle: Metz, 1944.* New York: Stein and Day, 1981.

Kennedy, John F. Message to Congress, 25 May 1961. Available at http://www.cs.umb.edu/jfklibrary/j052561.htm

Mabry, Robert L. et al. "United States Army Rangers in Somalia: An Analysis of Combat Casualties on an Urban Battlefield." *Journal of Special Operations Medicine* (fall 2001): 24–40.

MacArthur, Douglas. *Reminiscences.* New York: McGraw-Hill, 1964.

Medby, Jamison Jo and Russell W. Glenn. *Street Smart: Intelligence Preparation of the Battlefield for Urban Operations.* Santa Monica, CA: RAND, 2002.

Monmonier, Mark. *How to Lie with Maps.* Chicago: University of Chicago Press, 1996.

Montgomery, Thomas M. Interview. "Ambush in Mogasdishu." *Frontline,* WGBH Boston. 1998. Available at http://www.pbs.org/wgbh/pages/frontline/shows/ ambush/ interviews/montgomery.html.

Natsios, Andrew S. "Commander's Guidance: A Challenge of Complex Humanitarian Emergencies." *Parameters* (summer 1996): 50–66.

Natsios, Andrew S. "Food Through Force: Humanitarian Intervention and U.S. Policy." *The Washington Quarterly* (winter 1994): 136.

Nolan, Keith William. *The Battle for Saigon: Tet 1968.* New York: Pocket Books, 1996.

Oakley, Robert B. "An Envoy's Perspective." *JFQ Forum* (autumn 1993): 44–55.

Oliker, Olga. *Russia's Chechen Wars 1994–2000: Lessons From Urban Combat.* Santa Monica, CA: RAND, 2001.

Peters, Ralph. "Our Soldiers, Their Cities." *Parameters* (spring 1996): 43–50.

Peterson, John J. *Into the Cauldron.* Clinton, MD: Clavier House, 1973.

Senge, Peter M. *The Fifth Discipline: The Art and Practice of the Learning Organization*. New York: Currency Doubleday, 1990.

Shelton, H. Hugh and Timothy D. Vane. "Winning the Information War in Haiti." *Military Review* (November/December 1995): 3–9.

Shtemenko, S. M. *The Soviet General Staff at War, 1941-1945: Book One*. Moscow: Progress Publishers, 1981.

Smith, Rupert. "Wars in Our Time–A Survey of Recent and Continuing Conflicts." *World Defence Systems* (July 2001): 120–122.

Stanton, Shelby L. *Ten Corps in Korea, 1950*. Novato, CA: Presidio Press, 1989.

Thomas, Timothy L. "The Battle for Grozny: Deadly Classroom for Urban Combat." *Parameters* (summer 1999): 87–102.

Thomas, Timothy L. and Charles P. O'Hara. "Combat Stress in Chechnya: 'The Equal Opportunity Disorder'." *Army Medical Department Journal* (January-March 2000): 46–53.

United Nations Population Division. *World Urbanization Prospects: The 1999 Revision*. New York: United Nations, 1995.

US Agency for International Development, Bureau for Humanitarian Response, Office of Foreign Disaster Assistance. *Field Operations Guide for Disaster Assessment and Response*. Washington, DC: US Government Printing Office, 1994.

Yates, Lawrence A. "Military Stability and Support Operations: Analogies, Patterns and Recurring Themes." *Military Review* (July/August 1997): 51–61.

READINGS RECOMMENDED

These sources contain relevant supplemental information.

Alexander, John B. *Future War: Non-Lethal Weapons in Twenty-First-Century Warfare*. New York: St. Martin's Press, 1999.

Arnold, James R. *Tet Offensive 1968: Turning Point in Vietnam*. London: Osprey, 1990.

Ashworth, G. J. *The City and War*. London: Routledge, 1991.

Barker, A. J. *Dunkirk: The Great Escape*. New York: David McKay, 1977.

Breuer, William B. *Hitler's Fortress Cherbourg: The Conquest of a Bastion*. New York: Stein and Day, 1984.

Brown, Stephen D. "PSYOP in Operation Uphold Democracy." *Military Review* (September/October 1996): 57–76.

Bulloch, Gavin. "Military Doctrine and Counterinsurgency: A British Perspective." *Parameters* (summer 1996): 4–16.

Center for Army Lessons Learned (CALL) U.S. Army Combined Arms Command (CAC). *Urban Combat Operations: Tactics, Techniques, and Procedures, Newsletter No. 99-16*. Available at http://call.army.mil/.

Carell, Paul. *Stalingrad. The Defeat of the German 6th Army*. Atglen, PA: Schieffer Publishing, 1993.

Collins, J. Lawton. *Lightning Joe*. Novato, CA: Presidio Press, 1994.

Cordesman, Anthony H. and Abraham R. Wagner. *The Lessons of Modern War—Volume II: The Iran-Iraq War*. Boulder, CO: Westview Press, 1990.

Craig, William. *Enemy at the Gates: The Battle for Stalingrad*. New York: Reader's Digest Press, 1973.

Desch, Michael C., ed. *Soldiers in Cities: Military Operations on Urban Terrain*. Carlisle Barracks, PA: Strategic Studies Institute, 2001.

Dewar, Michael. *War in the Streets: The Story of Urban Combat from Calais to Khafji*. New York: Sterling Publishers, 1992.

Duffy, Christopher. *Fire & Stone: The Science of Fortress Warfare, 1660–1860*. Mechanicsburg, PA: Stackpole Books, 1996.

———. *The Fortress in the Age of Vauban and Frederick the Great, 1660–1789—Siege Warfare Volume I*. London: Routledge & Kegan Paul, 1985.

———. *Siege Warfare: The Fortress in the Early Modern World, 1494–1660*. London: Routledge & Kegan Paul, 1979.

Edwards, Sean J. A. *Mars Unmasked: The Changing Face of Urban Operations*. Santa Monica, CA: RAND, 2000.

Eikenberry, Karl W. "Take No Casualties." *Parameters* (summer 1996): 109–118.

Erickson, John. *The Road to Berlin: Continuing the History of Stalin's War with Germany*. Boulder, CO: Westview Press, 1983.

Eshel, David. *Mid-East Wars: The Lebanon War 1982*. Hod Hasharon, Israel: Eshel-Dramit, 1983.

Gerwehr, Scott and Russell W. Glenn. *The Art of Darkness: Deception and Urban Operations*. Santa Monica, CA: RAND, 2000.

Glenn, Russell W. *Combat in Hell: A Consideration of Constrained Urban Warfare*. Santa Monica, CA: RAND, 1996.

———. *Marching Under Darkening Skies: The American Military and Impending Urban Operations Threat*. Santa Monica, CA: RAND, 1998.

Glenn, Russell W., ed. *Capital Preservation: Preparing for Urban Operations in the Twenty-First Century: Proceedings of the RAND Arroyo-TRADOC-MCWL-OSD Urban Operations Conference, March 22-23, 2000*. Santa Monica, CA: RAND, 2001.

Glenn, Russell W. et al. *Denying the Widow-Maker, Summary of Proceedings, RAND-DDBL Conference on Military Operations on Urbanized Terrain*. Santa Monica, CA: RAND, 1999.

———. *Getting the Musicians of Mars on the Same Sheet of Music: Army Joint, Multinational, and Interagency C4ISR Interoperability*. Santa Monica, CA: RAND, 2000.

Glenn, Russell W. et al., eds. *The City's Many Faces: Proceedings of the RAND Arroyo-MCWL-J8 UWG Urban Operations Conference, April 13-14, 1999*. Santa Monica, CA: RAND, 2000.

Glenn, Russell W., Steven Hartman and Scott Gerwehr. *On the Shoulders of Atlas: Urban Combat Service Support Operations*. Santa Monica, CA: RAND, pending publication in 2003.

Gouré, Leon. *The Siege of Leningrad*. Stanford, CA: Stanford University Press, 1962.

Grau, Lester W. and Jacob W. Kipp. "Urban Combat: Confronting the Specter." *Military Review* (July/August 1999): 9–17.

Groves, John R. "Operations in Urban Environments." *Military Review* (July/August 1998): 30–39.

Hahn, Robert F., II and Bonnie Jezior. "Urban Warfare and the Urban Warfighter of 2025." *Parameters* (summer 1999): 74–86.

Heal, Sid. "Crowds, Mobs and Nonlethal Weapons." *Military Review* (March/April 2000): 45–50.

Heckstall-Smith, Anthony. *Tobruk: The Story of A Siege*. New York: W. W. Norton, 1959.

Harman, Nicholas. *Dunkirk: The Patriotic Myth*. New York: Simon and Schuster, 1980.

Hibbert, Christopher. *The Battle of Arnhem*. London: Batsford, 1962.

Hollis, Mark A. B. "Platoon Under Fire." *Infantry* (January-April 1998): 27–34.

House, John M. "The Enemy After Next." *Military Review* (March/April 1998): 22–28.

Jackson, W. G. F. *The Battle for Italy*. New York: Harper & Row, 1967.

Karslake, Basil. *1940 – The Last Act: The Story of the British Forces in France After Dunkirk*. Hamden, CT: Archon Books, 1979.

Kostof, Spiro. *The City Shaped: Urban Patterns and Meanings Through History*. London: Thames and Hudson, 1999.

———. *The City Assembled: The Elements of Urban Form Throughout History*. Boston: Little Brown, 1999.

Kuehl, Daniel T. "Airpower vs. Electricity: Electric Power As a Target for Strategic Air Operations." *Journal of Strategic Studies* (March 1995): 237–266.

Larson, Eric. *Casualties and Consensus: The Historical Role of Casualties in Domestic Support of U.S. Military Operations*. Santa Monica, CA: RAND, 1996.

Lieven, Anatol. *Chechnya: Tombstone of Russian Power*. New Haven: Yale University Press, 1998.

———. "The World Turned Upside Down: Military Lessons of the Chechen War." *Armed Forces Journal International* (August 1998): 40–43.

Linden, Eugene. "The Exploding Cities of the Developing World." *Foreign Affairs* (January/February 1996): 52–65.

McConnell, Malcolm. *Just Cause: The Real Story of America's High-Tech Invasion of Panama*. New York: St. Martin's Press, 1991.

McKee, Alexander. *Caen: Anvil of Victory*. New York: St. Martin's Press, 1964.

McLaurin, R. D. "Military Forces in Urban Antiterrorism." Aberdeen Proving Ground, MD: US Army Human Engineering Laboratory, 1989.

McLaurin, R. D. and R. Miller. "Urban Counterinsurgency: Case Studies and Implications for U.S. Military Forces." Aberdeen Proving Ground, MD: US Army Human Engineering Laboratory, 1989.

Mendel, William W. *Combat in Cities: The LA Riots and Operation Rio*. Fort Leavenworth, KS: Foreign Military Studies Office, 1996.

Mercier, Charles E. "Terrorists, WMD, and the US Army Reserve." *Parameters* (autumn 1997): 98–118.

Milton, T. R. "Urban Operations: Future War." *Military Review* (February 1994): 37–46.

Moskin, J. Robert. *Among Lions: The Battle for Jerusalem, June 5–7, 1967*. New York: Ballantine Books, 1982.

Northam, Ray M. *Urban Geography*. New York: Wiley, 1975.

Parkerson, John Embry, Jr. "United States Compliance with Humanitarian Law Respecting Civilians During Operation Just Cause." *Military Law Review* (1991): 31–140.

Peters, Ralph. "The Culture of Future Conflict." *Parameters (*winter 1995-96): 18–27.

———. "The Future of Armed Warfare." *Parameters* (autumn 1997): 50–59.

———. "The Human Terrain of Urban Operations." *Parameters* (spring 2000): 4–12.

Rosenau, William G. "'Every Room is a New Battle': The Lessons of Modern Urban Warfare." *Studies in Conflict and Terrorism* (October–December 1997): 371–394.

Rosenberger, Leif R. "The Strategic Importance of the World Food Supply." *Parameters* (spring 1997): 84–105.

Scalard, Douglas. "People of Whom We Know Nothing: When Doctrine Isn't Enough." *Military Review* (July/August 1997): 4–11.

Scales, Robert H. "The Indirect Approach: How US Military Forces Can Avoid the Pitfalls of Future Urban Warfare." *Armed Forces Journal International* (October 1998): 68–74.

Siegel, Adam B. *Eastern Exit: The Noncombatant Evacuation Operation (NEO) from Mogadishu, Somalia, in January 1991*. Alexandria, VA: Center for Naval Analyses, 1992.

Spiller, Roger J. *Sharp Corners: Urban Operations at Century's End.* Fort Leavenworth, KS: Combat Studies Institute, 2001.

Stech, Frank J. "Winning CNN Wars." *Parameters* (autumn 1994): 37–56.

Stevenson, Jonathan. *Losing Mogadishu: Testing U.S. Policy in Somalia.* Annapolis, MD: Naval Institute Press, 1995.

Taw, Jennifer Morrison. *Operation Just Cause: Lessons for Operations Other Than War.* Santa Monica, CA: RAND, 1996.

Taw, Jennifer Morrison and Bruce Hoffman. *The Urbanization of Insurgency: The Potential Challenge to U.S. Army Operations.* Santa Monica, CA: RAND, 1994.

Tully, Andrew. *Berlin: Story of a Battle.* Westport, CT: Greenwood Press, 1977.

Vick, Alan et al. *Aerospace Operations in Urban Environments: Exploring New Concepts.* Santa Monica, CA: RAND, 2000.

Watson, Bruce Allen. *Sieges: A Comparative Study.* Westport, CT: Praeger, 1993.

Weigley, Russell. *Eisenhower's Lieutenants.* Bloomington, IN: Indiana University Press, 1981.

Werstein, Irving. *The Battle of Aachen.* New York: Thomas Y. Crowell Company, 1962.

Whiting, Charles. *Bloody Aachen.* New York: Stein and Day, 1976.

Wintle, Justin, ed. *The Dictionary of War Quotations.* New York: The Free Press, 1989.

Wykes, Alan. *The Siege of Leningrad.* New York: Ballentine Books Inc., 1968.

Index

Entries are by paragraph number unless figure (fig.) or page (pg.) is specified.

D

E–F

G–H

J–K–L

M

Q–R

S

U–V

W-X-Y-Z

FM 3-06 (FM 90-10)
1 JUNE 2003

By Order of the Secretary of the Army:

ERIC K. SHINSEKI
General, United States Army
Chief of Staff

Official:

Joel B. Hudson

JOEL B. HUDSON
Administrative Assistant to the
Secretary of the Army
0312903

DISTRIBUTION:

Active Army, Army National Guard, and U. S. Army Reserve: To be distributed in accordance with the initial distribution number 111233, requirements for FM 3-06.

PIN: 080805-000

Made in the USA
Monee, IL
04 November 2026

67384911R00193